Diesel Engines
for Automobiles
and Small Trucks

Diesel Engines for Automobiles and Small Trucks

Tom Weathers, Jr.

Claud Hunter

 Reston Publishing Company, Inc.
A Prentice-Hall Company
Reston, Virginia

Library of Congress Cataloging in Publication Data

Weathers, Tom.
   Diesel engines for automobiles and small trucks.

   Includes index.
   1. Automobiles—Motors (Diesel)   2. Automobiles—
Motors (Diesel)—Maintenance and repair.
I. Hunter, Claud C., joint author.   II. Title.
TL229.D5W43   629.2'506   80-29478
ISBN 0-8359-1288-4

© 1981 by
Reston Publishing Company, Inc.
A Prentice-Hall Company
Reston, Virginia 22090

10   9   8   7   6   5   4   3   2   1

Printed in the United States of America

Contents

v

Chapter 8   Diesel Exhaust Systems   101

Chapter 9   Diesel Fuel and Injection Systems   107

Chapter 10   Diesel Cooling Systems   139

Chapter 11   Diesel Lubricating System   149

Preface

A revolution is taking place in the transportation industry. Almost every year, something radically different comes along. Some developments don't lead anywhere or have only limited use, as was the case with the Wankel rotary engine. Considered to be a potential replacement for the conventional four-stroke engine in the early 1970s, rotaries today are only used in a few specialty vehicles. They simply didn't offer sufficient improvement in the areas that count: economy and pollution control.

Developments that do satisfy these requirements often become widely accepted. Automobile and small truck diesels are an example. Although not likely to completely replace gasoline engines, diesels do offer substantial advantages in economy and pollution control. Judging from the number of diesel-powered vehicles on the road and from the financial commitment made by many large companies to their use and development, diesels appear to have carved out a definite niche in the personal transportation market. Unlike the Wankel engine, automotive and small truck diesels are here to stay—and in significant numbers.

Where does that leave you? If you are either studying to be a mechanic or have experience as a mechanic, the odds are that you will be faced with repairing a diesel engine some day. If you are the owner of a diesel-powered vehicle, it is certain that your engine will need service at some point.

The authors of this book believe that, before you can fix anything, you need to know how it works. Even before consulting a manufacturer's shop manual or a specialized troubleshooting guide, you should have a grounding in the fundamentals.

Providing such groundwork is the object of this book. First, the operation of the diesel engine is described. Then basic diesel engine systems, electrical systems, tune-up and troubleshooting procedures, major service operations, and tools and equipment are covered. Since mechanics or owners may encounter more than one type of diesel, the theoretical discussions are illustrated by examples from major diesel manufacturers. Also, since most people are already somewhat familiar with gasoline engines, diesel and gasoline engine operation are compared throughout the book to help clarify the ideas presented.

Unless you are taking a formal course of study, the best way to approach this book is to first scan the introductory paragraphs of each chapter. This will give you an overall idea of what the book is about. If you already know something about gasoline engines, it will also help you to discover that many aspects of gasoline and diesel engines are the same. You'll realize that learning about diesels may not be as difficult as you imagined.

The next step, obviously, is to read the book in detail from start to finish. Make sure you understand each chapter before going on to the next so that the facts will accumulate in a "building block" manner—one securely based on the others.

After finishing the book, you won't be a diesel mechanic. A lot of hands-on practice will be required for that. However, you will be properly prepared. You will have the broad base of knowledge required to deal with specialized shop manuals. Most important, you will have the groundwork for handling the unique problems that no instructor or author can predict but that everyone knows will come up.

The authors wish to gratefully acknowledge the help of the following companies: Nissan Diesel Motors Ltd./Marubeni America Corporation; Mercedes-Benz; Volkswagen; Oldsmobile Division, General Motors Corporation; Gould, Inc; Peugeot; Robert Bosch Corporation; Chrysler Corporation; Prentice-Hall, Inc; New Britain Hand Tools; DeKoven Manufacturing Company, Ammco Tools, Inc.; Kleer-Flo Company; Jenny Division, Homestead Industries; Black & Decker; and Balcrank Products Division, Wheelabrater-Frye, Inc.

Tom Weathers, Jr.
Claud Hunter

CHAPTER 1

Introduction to Diesel Engines

GENERAL

This chapter outlines the history of diesel engines and notes some advantages and disadvantages of diesel engines as they relate to gasoline engines. Some applications of diesel engines are also given. The next chapter examines the operating principles of diesels.

HISTORY

Credit for inventing the diesel engine concept goes to the German engineer Doctor Rudolf Diesel. He first presented his concept for a "rational heat engine" in 1892. That was 20 years after the first gasoline engine was built—in 1863 by a Frenchman named Lenoir.

Doctor Diesel originally tried to build an engine using coal dust as a fuel. It was readily available and cheap. And it seemed like a good idea to Doctor Diesel and his backers, the wealthy German industrialist, Baron Friedrich Von Krupp, and a large company called Machine-Fabrick Augsburg Nurnberg (M.A.N.). Unfortunately, the idea didn't work. It wasn't until 1897 that Doctor Diesel, aided by another scientist named Lauster and the engineers of the M.A.N. Company, produced a working liquid fuel engine.

Doctor Diesel disappeared mysteriously in 1913; however, his ideas had already taken hold. Adolphus Bush of beer fame obtained a license to manufacture and sell diesel engines in the United States and Canada. The first diesel engine to go into service anywhere in the world was an industrial engine used in one of the Bush breweries.

FIGURE 1–1.   A very early diesel engine (*Courtesy of Prentice-Hall, Inc.*).

The first diesel powered truck was produced in 1923 by Daimler-Benz. The first series production diesel automobile was introduced at the 1936 Berlin Auto Show by Mercedes. This car was primarily intended as a taxi. It had a 2.6 liter, four-cylinder engine giving a top speed of about 60 mph. About 2,000 were built before World War II began.

FIGURE 1–2.   The first Daimler-Benz diesel car (*Courtesy of Mercedes-Benz*).

APPLICATIONS

Diesel engines have been (and still are to a degree) used primarily for heavy duty, commercial applications. They are used in large trucks, railroad locomotives, and ships as well as in many stationary applications. Automotive diesels until recently were only produced by a few companies. The reason for the commercial slant of diesels has been primarily a matter of economics. The current interest in diesels for passenger cars is also a matter of economics, coupled with a concern for the environment.

To see why this is so, we need to look at the advantages and disadvantages of diesel and gasoline engines.

FIGURE 1-3.   Some other kinds of diesel applications (*Courtesy of Reston Publishing Company, Inc.*).

ADVANTAGES OF DIESELS

Economy

One of the main advantages of diesel engines is fuel economy. A diesel generally uses less fuel than the same size gasoline engine in comparable vehicles. According to Mercedes-Benz, in 1974 their 240 series diesel car delivered 70% better fuel economy than other vehicles in the 3,500 pound class.

One reason for a diesel's fuel economy is the thermal efficiency of the diesel fuel compared to gasoline. Thermal efficiency is a measure of how much of the fuel's energy is converted to useful, mechanical work versus the amount of energy wasted as exhaust or radiated heat. As noted in Figure 1-4 a typical diesel passenger car engine delivers a higher percentage of thermal efficiency than gasoline engines at all load and speed ranges.

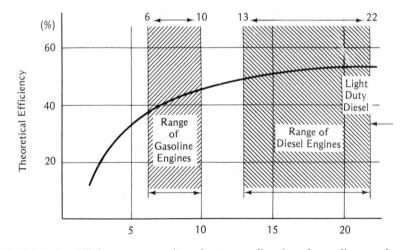

FIGURE 1-4. Efficiency comparison between diesel and gasoline engines.

Another reason for superior diesel fuel economy is the fact that diesel fuel usually costs less than premium or low-lead gasoline. Diesel fuel takes fewer refining steps and is cheaper to produce. (Although enough new passenger diesels might overtax the oil companies' ability to produce diesel fuel and by the process of supply and demand drive up the price.)

Low Emissions

Diesel engines produce lower levels of certain kinds of pollutants than gasoline engines. Because the peak combustion temperatures of diesel engines are lower than gasoline engines, NO_x (nitric oxide) formation is less. Also, because diesels always operate with an excess of air in the combustion chamber, there is plenty of oxygen to burn away most excess hydrocarbons and to reduce most carbon monoxide (a poison) to carbon dioxide.

However, diesel engines do produce more particulate or solid pollutants than gasoline engines. This is the smoke commonly seen coming from diesel exhaust pipes. Until recently, particulate emission, although not particularly attractive, was not considered a significant health hazard. That view is being challenged by environmental groups and the government. At the time of this writing, the outcome of the investigation is not clear, but it is possible that diesel manufacturers will have to reduce the solid emissions produced by their engines.

Reduced Maintenance

Since diesel engines do not have an ignition system as such, there is no need for regular ignition system service. This was a decided advantage when most passenger car engines had points-type ignition systems that required replacement of the points, condenser, and spark plugs at regular intervals. However, the advantage is not so great now that most passenger cars use electronic ignition systems. The points and condenser have been eliminated leaving only the spark plugs—and they do not require replacement as often as before. Also, diesel engines usually require more frequent oil and filter changes than gasoline engines. So, reduced maintenance is a questionable advantage.

Reliability

This is also a questionable advantage in passenger car engines. It is true that most truck and industrial diesel engines remain in service longer than most passenger car gasoline engines. It is also true that certain manufacturers of passenger car diesels have a reputation for building durable engines. And it is true that a diesel engine has to be stronger than a gasoline engine to withstand the increased combustion chamber pressures and temperatures. However, it does not follow that a diesel engine must last longer than a gasoline engine. It depends on the manufacturer's goals. When selling to industrial and commercial markets, it is important to have a long lasting product. So, the manufacturers of commercial diesels overbuilt their engines in comparison to gasoline engines. The passenger car engines in service until recently were also "overbuilt" because the same engines that went into private use were also sold for commercial applications (like taxi and fleet use) where extreme durability is required.

It remains to be seen whether the same philosophy will hold for the new breed of passenger car diesels. However, even if it doesn't and passenger car diesels last no longer than comparable gasoline engines, it doesn't mean that the diesel engines will not give long, trouble-free service in passenger car use. And the advantages of reduced pollution, improved fuel economy, and somewhat improved maintenance will remain regardless.

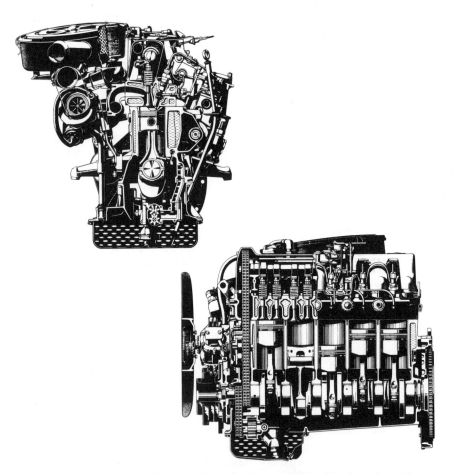

FIGURE 1-5.   Cutaway view of Mercedes-Benz 300SD diesel engine (*Courtesy of Mercedes-Benz*).

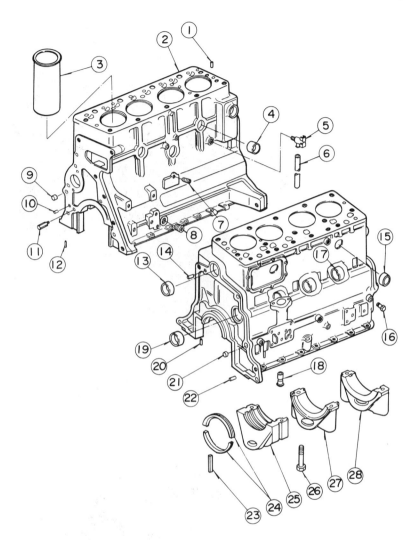

1	Dowel pin	11	Oil jet	21	Oil seal
2	Cylinder block	12	Plug	22	Main bearing cap (No. 4)
3	Cylinder liner	13	Straight pin	23	Main bearing cap (No. 3)
4	Plug	14	Plug	24	Main bearing cap (No. 2)
5	Water drain cock	15	Plug	25	Main bearing cap (No. 1)
6	Vinyl hose	16	Straight pin	26	Bushing
7	Plug	17	Plug	27	Plug
8	Plug	18	Straight pin	28	Plug
9	Plug	19	Oil seal	29	Camshaft bushing
10	Dowel pin	20	Main bearing cap bolt		

FIGURE 1-6. A Nissan diesel engine block (*Courtesy of Nissan Diesel Motors Ltd./Marubeni America Corporation*).

FIGURE 1–7.   A Mercedes-Benz diesel engine (*Courtesy of Mercedes-Benz*).

FIGURE 1–8.   A Peugeot diesel engine (*Courtesy of Peugeot*).

DISADVANTAGES OF DIESELS

Lower Output and Torque

Given identical displacement engines, a diesel will only produce 65% of the horsepower and 80% of the torque of a gasoline engine. Looking at it another way, a theoretical 2-liter gasoline engine produces the same horsepower as a 3-liter diesel and the same torque as a 2.5-liter diesel.

Higher First Cost

Because diesel engines need to be larger to give the same horsepower and torque as a gasoline engine and because they need to be stiffer and stronger to withstand the additional pressures and temperatures, diesel engines usually cost more to make. However, in high mileage applications (taxi, truck, etc.) the extra cost is usually offset by better fuel economy and reduced maintenance.

Higher Weight

Because diesels are stiffer and stronger than comparable gasoline engines, they usually weigh more.

Increased Noise

Owing to the operating characteristics of diesels, they are usually noisier than gasoline engines, particularly just after cold starts. This noise, which is actually a kind of detonation or spark knock, is not considered harmful.

Smoke

Diesels produce more particulate or visible emission than gasoline engines. However, it is possible that government regulation will force manufacturers to eliminate this objectionable feature in future products.

SUMMARY

As we've seen, diesel engines are better than gasoline engines in some respects and worse in others. Until recently, the disadvantages for passenger car use have largely outweighed the advantages. However, in an era of scarce fuel and polluted air, the advantages begin to seem more important. It is not likely that the diesel will ever replace the conventional gasoline engine—it is more likely that some other kind of powerplant will replace both. However, until that distant day, it does seem probable that more and more private passenger cars will have diesel engines.

FIGURE 1–9. An experimental diesel racing car (*Courtesy of Mercedes-Benz*).

CHAPTER 2

Principles of Operation

GENERAL

This chapter introduces the operating principles of diesel engines. Diesel and gasoline engines are compared, and four- and two-stroke cycles of operation explained.

DIESELS COMPARED TO GASOLINE ENGINES

Diesel engines are like gasoline engines in many ways. Both operate by internal combustion, e.g., by burning fuel inside a closed space. Both employ a four-stroke, two-stroke, or rotary system of operation. Both use many of the same mechanical components: pistons (except for rotaries), cylinders (with the same exception), crankshafts, etc.

The differences are due to the nature of the fuels used. Gasoline is a high volatility fuel; it evaporates readily to become a vapor. Diesel fuel is less volatile. These differences dictate different approaches to combining air and fuel into a combustible mixture and to igniting the mixture once it is combined.

REVIEW OF
GASOLINE POWERED ENGINES

Any fuel, to burn properly, must be mixed with air. Reliable ignition and satisfactory air/fuel distribution are achieved in gasoline engines by mixing air and fuel before they go into the combustion chamber. That gives the gasoline time for at least partial evaporation so the individual gasoline molecules can get close to the oxygen molecules required for combustion. This intimate mixing takes place in the relatively low pressure regions of the carburetor (if the engine is so equipped) and the intake manifold.

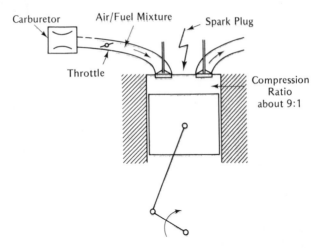

FIGURE 2-1.   Schematic of a gasoline engine.

Gasoline engines mix air and fuel in a more-or-less stoichiometric ratio. This refers to the amount of air that must be present to completely burn a given amount of fuel. It takes about 14 pounds of air to consume one pound of fuel. Therefore the carburetor or fuel injector of a gasoline engine must supply about one part of fuel for every 14 parts of air that pass the throttle plate.

The operating range of gasoline engines is about 20% plus or minus the theoretical stoichiometric air/fuel ratio. The mixture must stay within this range for the engine to operate properly.

If the engine breathes too much air, it operates lean. Generally this is most likely to occur at part load when the engine is running slowly even though the throttle is open. The engine misfires and produces unburned hydrocarbons.

If the engine takes in less air than it needs to stay within the stoichiometric operating range, the result is an excessively rich mixture. Fuel consumption goes up and carbon monoxide formation increases—the latter taking place because there is not enough air to form carbon dioxide. Figure 2-2 notes the chemical relationships in ideal combustion when there is no fuel or pollutants left over. Figure 2-3 describes in graphic form the operating range of a normal gasoline engine.

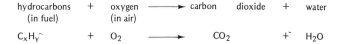

FIGURE 2-2. Formula showing the complete nature of the diesel combustion process.

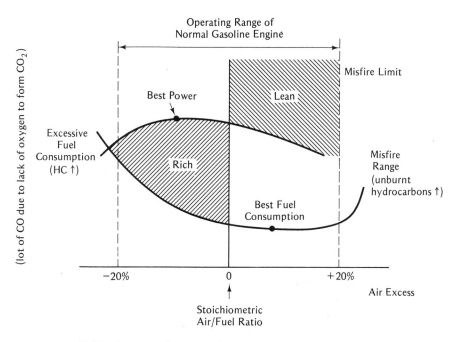

FIGURE 2-3. The operating range of gasoline engines.

DIESEL ENGINE DIFFERENCES

Like gasoline engines, diesels must mix air and fuel into a combustible mixture. Individual molecules of fuel must enter into close proximity with molecules of oxygen for combustion to take place. However, the ways of achieving this mixture and igniting it are different.

First of all, the air drawn into the combustion chamber of a diesel does not contain any fuel. Nor does a diesel generally have an air valve or throttle. The engine takes in as much air as the restrictions in the air flow passages and the speed and displacement of the engine will allow.

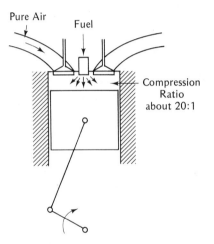

FIGURE 2-4.   Schematic of a diesel engine.

Viewed in terms of the stoichiometric ratio, diesels draw in as much as 600% more air than is needed for combustion at idle speeds and as much as 25% extra air at full throttle. Figure 2-5 compares the operating range of diesels versus gasoline engines.

Diesels mix fuel and air directly in the combustion chamber. The speed of the engine is governed by the amount of fuel introduced. Without some sort of governor, the engine would run away, producing as many as 2,000 power impulses per second.

High pressure injectors are used to spray the diesel fuel into the combustion chamber. Unlike gasoline engines, the injection and subsequent air/fuel mixing occur late in the compression stroke. And unlike gasoline engines, diesels operate at much higher compression ratios, in the range of 14:1 to 24:1. This means the air pressure in the combustion chamber is 35 to 45 times greater than normal atmospheric pressure (14.5 pounds per square inch). To obtain good penetration of fuel into the pressurized air and to achieve good atomization and fuel distribution, the velocity of the air droplets has to be proportionally higher. Diesel fuel injection systems therefore always operate at high pressures, in the range of 250–350 atmospheres. Adding the pressure created by the combustion process with the pressure due to the compression stroke results in peak combustion chamber pressures of 55 to 75 atmospheres.

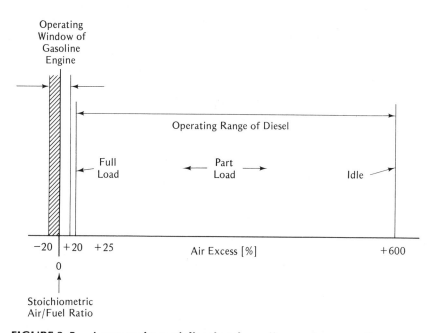

FIGURE 2–5.   A comparison of diesel and gasoline engine operating ranges.

Diesels operate at such high pressures in the combustion chamber because of the way the fuel is ignited. As air is compressed, it becomes hotter. By the time fuel is injected into the combustion chamber of a diesel engine, the air is so hot that the fuel self-ignites without the need for a spark plug (although glow plugs and other means are sometimes used to help start cold engines). That is why diesels are often called self-ignited engines.

The increased compression ratio also has another effect. Higher compression ratios result in greater thermal efficiency. As noted in the previous chapter, thermal efficiency is a measure of how well an engine converts chemical energy into mechanical force. In other words, if the thermal efficiency of an engine is 40%, then 40% of the fuel it uses is converted into work and the other 60% is wasted as heat lost from the exhaust pipe or radiated into the atmosphere via the radiator.

However, even though higher compression ratios provide self-ignition and better thermal efficiency, the increased pressures and stresses require diesels to be stronger and stiffer than comparable gasoline engines.

There are also other differences: many gasoline engines have carburetors to mix air and fuel; diesels don't need them. Gasoline engines require elaborate ignition systems including spark plugs, coils, distributors, etc. As we saw, diesel engines are self-igniting and don't require any of these components. Many of the other differences between diesels and gasoline engines are due to the ways the engines have been used: diesels primarily for heavy duty, industrial applications; gasoline engines for automotive and lighter duty use. However, as we saw in the last chapter and as we'll confirm again in this book on automotive diesels, the old distinctions are becoming blurred. Figure 2–6 summarizes the major differences between diesel and gasoline engines.

FOUR-STROKE CYCLE OPERATION

Like a four-stroke gasoline engine, a four-stroke diesel engine has an intake, compression, power, and exhaust stroke for each operating cycle. Like a gasoline engine, each stroke corresponds to one complete up or down movement of the piston. And, like a gasoline engine, the resulting four strokes equal two complete crankshaft revolutions, or 720 degrees.

The differences relate to the differences between diesel fuel and gasoline and how each must be handled so its chemical energy can be converted into useful work to push the piston down and rotate the crankshaft. The following paragraphs and illustrations describe the four strokes in a typical diesel operating cycle.

Intake

The intake stroke begins just before the piston reciprocates past top dead center and begins its downward movement in the cylinder. Relating piston movement to crankshaft rotation and considering top dead center (TDC) equal to 0 degrees of crankshaft rotation, the intake stroke begins at about 20–30 degrees before top dead center (BTDC). This is the point when the intake valve begins to open. A camshaft lobe

Comparison of Characteristic Features of Gasoline and Diesel Engines

engine type / category	Gasoline (Otto-engine, SI-engine	Diesel (CI-engine)
Inventor	Nikolaus A. Otto (1832 – 1891)	Rudolf Diesel (1857 – 1913)
fuel	high volatility	low volatility
fuel feed	early (in case of injection: low pressure)	late (always injected, high pressure injection)
inlet	air/fuel mixture	air only
compression ratio	of mixture 8 – 10	of air 14 – 22
peak temperature and pressure	medium (35 – 45 at)	high (55 – 75 at)
ignition	spark	self
air/fuel ratio	stoichiometric ± 20%	lean + 30% to 600%
mixture	generally fairly homogeneous	fairly stratified due to concentrated late injection
load control	by throttle, controlling amount of mixture (quantity control)	by controlling amount of injected fuel (quality control)

FIGURE 2-6.   A summary of the major differences between diesel and gasoline engines.

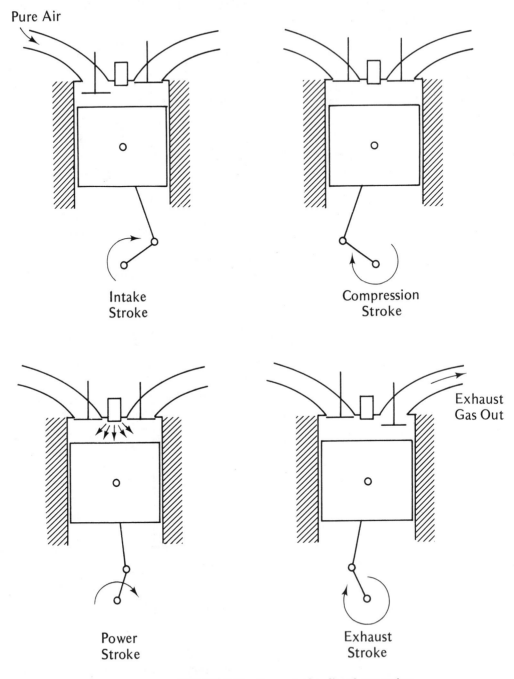

FIGURE 2-7.   **Four-stroke diesel operation.**

starts taking up the slack (play) in the valve train components between the lobe and the intake valve.

As the intake valve opens, the piston goes past TDC and begins to move down in the cylinder. A low pressure region is created inside the cylinder. Air at normal atmospheric pressure outside the engine is pushed into the vacant, low pressure area inside the cylinder. Unlike a gasoline engine, this air is not mixed with fuel.

The intake stroke ends shortly after the piston reciprocates past the bottom point of its travel and begins to move back up in the cylinder. Viewed in terms of crankshaft rotation, the intake stroke ends at about 30–40 degrees after bottom dead center (ABDC) when the intake valve begins to close.

Compression Stroke

The end of the intake stroke, when the intake valve closes to seal off the combustion chamber, signals the beginning of the compression stroke. It continues for the next 150 degrees or so of crankshaft rotation, until the piston reaches the top of the cylinder. Depending on the compression ratio of the engine, the volume in the cylinder is reduced as much as 14 to 24 times. The air drawn in during the intake stroke is squeezed tighter and tighter, which forces the vibrating molecules of air closer and closer together. As a result, the temperature of the air in the combustion chamber goes up and up. By the time the piston has reached the top of the cylinder, the air temperature is as much as 1200 to 1300° F. By the same token, the air pressure has also increased, up to 700–800 pounds per square inch.

Stroke

Although the power stroke doesn't begin until after TDC when the piston is actively pushed back down into the cylinder, the fuel required for the power stroke is injected late in the compression stroke. The high pressure fuel injectors open up at about 20–30 degrees BTDC and continue to spray fuel into the combustion chamber until TDC or a little longer. This is necessary to give the fuel time to vaporize and start burning. Usually the combustion process occurs in several stages:

First Stage. This is called the ignition delay period. Fuel injection begins, but the fuel isn't ignited yet. The hot air surrounding individual fuel droplets causes the fuel to evaporate into a cloud of combustible vapor. The purpose of the delay period is to begin to bring the fuel and air into the close proximity required for combustion so the next ignition stage will proceed smoothly. The delay must be long enough to achieve the proper distribution, but not so long that ignition, when it does occur, results in a too-rapid pressure rise. Factors affecting the delay period include air flow patterns in the combustion chamber (due to combustion chamber design), fuel droplet size, injection pressures, spray patterns, etc.

Second Stage. As compression and fuel injection continue, some of the fuel vapor injected in the first stage starts to burn. As the vapor surrounding individual droplets is consumed in the flames, the interior of the droplets evaporates to produce more fuel vapor, which is in turn burned away. The second stage begins shortly before the piston reaches TDC and continues for a few degrees of rotation afterward. Although temperature and pressures will inevitably go up in this period, designers try to keep the increases smooth and relatively low. Otherwise, power delivery will be uneven.

Third Stage. This stage occurs as the last droplets of fuel are injected into the combustion chamber. Temperatures are at their highest and the fuel ignites almost instantly after it is injected. After all (or nearly all) of the fuel is burned, the power stroke begins. The piston is pushed down in the cylinder by the expanding gases produced in the combustion process. Nearly constant pressure is exerted on the top of the piston until about 60–70 degrees after top dead center (ATDC). That is the point when the combustion products are no longer able to occupy a volume larger than the combustion chamber. This is also roughly the point when the angle between the piston connecting rod and crankshaft is greatest, thus giving the greatest mechanical advantage and exerting the maximum force on the crankshaft.

Exhaust Stroke

The exhaust valve begins to open before the piston reaches BDC, at about 45–55 degrees before bottom dead center (BBDC). After the piston reciprocates past bottom dead center, the combustion by-products are pushed out the exhaust port as the piston moves back up in the cylinder. However, the piston's speed up and down in the cylinder is not constant. It accelerates from no movement at one end of the cylinder until it reaches a certain speed, then it decelerates back to no movement at the other end of the cylinder. So, as the piston approaches the end of the exhaust stroke, it is slowing down. This results in air movement that causes the pressure in the combustion chamber to be slightly less than the pressure of the outside atmosphere. To get the remaining exhaust gases out of the combustion chamber, the opening of the intake valve is timed to overlap the closing of the exhaust valve. In other words, both valves are open at the same time. Air at normal atmospheric pressure coming in through the intake valve pushes exhaust gases out the exhaust valve. This is called scavenging.

TWO-STROKE CYCLE

There are very few (if any) two-stroke diesels in automotive use. However, considering the rapid pace of automotive diesel development, this situation might change over the next few years. So, the following paragraphs will briefly describe the principles of two-stroke diesel operation.

The same events must take place in two-stroke diesels as in four. Air must be drawn into the combustion chamber and compressed; fuel must be injected and burned; exhaust gases must be removed from the engine. However, two-stroke diesels combine all these operations into one up and one down movement of the piston. Viewed in terms of crankshaft rotation, everything takes place in 360 degrees (as opposed to 720 degrees in a four-stroke).

Intake

Air for two-stroke diesels is supplied by a blower or compressor driven off the crankshaft. The blower forces air into a manifold chamber at about 2 or 7 pounds per square inch. Ports connect the manifold to the walls of the cylinder. As the piston moves down in the cylinder, it uncovers the ports from the intake manifold. This is like opening an intake valve. Pressurized air flows into the cylinder. In most two-stroke diesels, the exhaust valves are still open at this time, allowing the pressurized air to sweep the exhaust out of the combustion chamber. That is why the blower is used—to scavenge exhaust gases and to fill the cylinders with air; not to act as a supercharger (although superchargers are used too, as noted in later chapters). The intake portion of the cycle continues as the piston moves down in the cylinder. It ends after the piston moves past bottom dead center and goes back up in the cylinder to close off the intake ports.

Compression

Compression, as in a four-stroke engine, takes place when both the exhaust valve and intake ports are closed and while the piston moves up in the cylinder. And as in a four-stroke engine, the effect is the same.

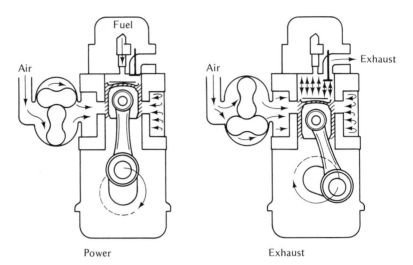

Power   Exhaust

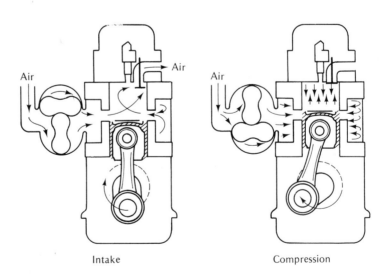

Intake   Compression

FIGURE 2-8. Two-stroke diesel operation.

Air is heated enough so that fuel sprayed into the combustion chamber near the end of the compression stroke ignites as a result of the heat.

Power

After the fuel has burned, the expanding gases push the piston back down in the cylinder. This is the portion of the cycle that delivers power to the piston, the crankshaft, and on to the driving wheels.

Exhaust

After the piston has traveled about 90 degrees, or halfway down in the cylinder in the power stroke, the exhaust valve opens. Not long after that, at about 130 degrees, the intake ports are uncovered. Both remain open until the piston starts back up in the cylinder, until about 40–50 degrees after bottom dead center.

It is by overlapping the exhaust and intake functions that two-stroke engines are able to do everything in 360 degrees instead of 720 degrees. In effect, there are only power and compression strokes with intake and exhaust taking place nearly simultaneously during the last part of the power stroke and the first part of the compression stroke.

But, you might ask, how can fresh intake air and exhaust gases occupy the combustion chamber at the same time? This is how: The exhaust valve opens a little before the intake ports, so some of the exhaust gases are removed before any fresh air is forced in the cylinder. Then, after the intake ports open, the swirling action of the pressurized air from the blower gradually replaces the exhaust gases with clean air. By the time the intake ports close, the air in the combustion chamber is virtually free of combustion by-products.

Now, you might ask, does a two-stroke engine develop twice as much horsepower as a four-stroke? It seems logical since the two-stroke has twice as many power impulses. Unfortunately, it doesn't work that way. Part of compression and power strokes must be sacrificed to give time for intake and exhaust functions, which as we noted before, occupy the last part of one stroke and the first part of the other.

SUMMARY

We've now examined the basic operating principles of diesel engines and studied four- and two-stroke operation. In the next chapter, we'll look at some of the major components of diesel engines.

———————————————————

CHAPTER 3

Engine Components

GENERAL

In previous chapters we briefly examined the history of diesel engines. We noted some applications as well as some disadvantages and advantages and studied the operating principles of four- and two-stroke cycle diesel engines. In this chapter, we'll look at the main components of diesel engines and point out the principal differences between these components and their gasoline engine counterparts. The items examined include:

1. Engine Blocks
2. Cylinder Heads
3. Crankshafts
4. Engine Bearings
5. Pistons, Rings, and Rods
6. Camshafts, Valves, and Valve Trains

ENGINE BLOCK

The block is the largest single part of the engine and one of the most important. It provides space for the cylinders, the crankshaft, and in some instances, the camshaft and various other components. The block must be strong enough to contain the combustion forces so that they can be converted into useful work, but it cannot be excessively heavy. Its design determines the configuration of the engine, in-line or vee, large or small. In short, the block is the basic framework of the engine.

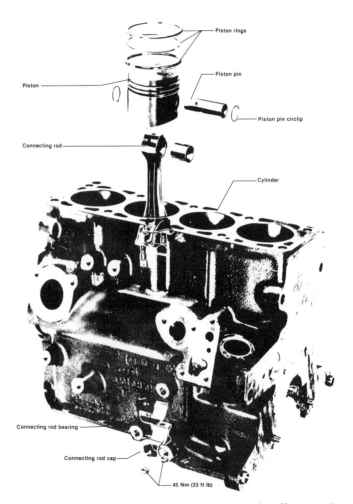

FIGURE 3-1.   An engine block (*Courtesy of Volkswagen*).

FIGURE 3–2. An engine block and crankshaft (*Courtesy of Nissan Diesel Motors Ltd./Marubeni America Corporation*).

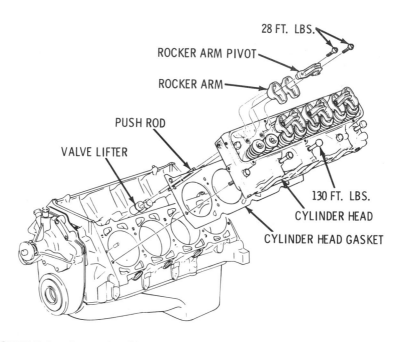

FIGURE 3–3. An engine block and cylinder head (*Courtesy of Oldsmobile Division, General Motors Corporation*).

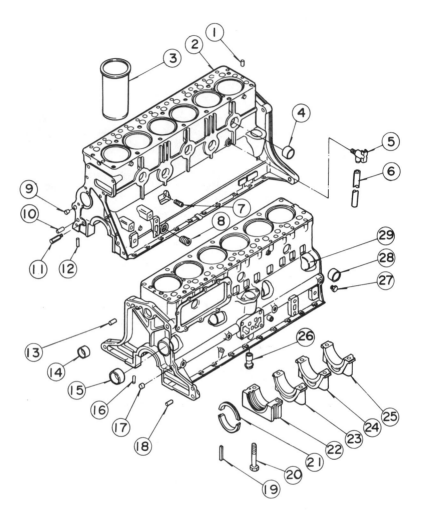

1	Dowel pin	11	Oil jet	21	Oil seal
2	Cylinder block	12	Plug	22	Main bearing cap (No. 4)
3	Cylinder liner	13	Straight pin	23	Main bearing cap (No. 3)
4	Plug	14	Plug	24	Main bearing cap (No. 2)
5	Water drain cock	15	Plug	25	Main bearing cap (No. 1)
6	Vinyl hose	16	Straight pin	26	Bushing
7	Plug	17	Plug	27	Plug
8	Plug	18	Straight pin	28	Plug
9	Plug	19	Oil seal	29	Camshaft bushing
10	Dowel pin	20	Main bearing cap bolt		

FIGURE 3–4.   An engine block and related parts (*Courtesy of Nissan Diesel Motors Ltd./Marubeni America Corporation*).

Casting

Most diesel engine blocks are cast in one piece from grey iron, or in some cases, from aluminum. Cast iron blocks are made by pouring molten iron into a sand mold. The cylinders and passages for liquid coolant are formed by installing sand cones into the mold.

Machining

After the block has cooled, it is removed from the mold and cleaned. Then it is machined into precise shape. The holes for attaching other engine parts are drilled and tapped; mating surfaces are milled and planed; cylinders are bored to exact diameters; and camshaft bearing openings (if the camshaft is located in the block) are bored and line-reamed. Plus, the block is machined to accommodate the valve lifters (if the camshaft is in the block) and the oil galleries and passageways. (See Figure 3–4).

Crankshaft Accommodations

The lower part of the block extends down to the approximate center line of the crankshaft. The main bearing caps (which position the crankshaft in the block) are cast separately, machined, and then installed on the block. After that, both the bearings caps and the block are line-reamed or bored to accommodate the crankshaft main bearings. The number of main bearings is determined by the number of cylinders and the design requirements of the engine.

Crankcase

The outside lower area of the diesel engine block is machined to accept the oil pan, which is attached to the block by cap screws. The area encased by the lower part of the block and the oil pan form the crankcase, literally, the case for the crank.

Strength Considerations

Any block, whether it is for a gasoline or diesel engine, requires stiffening and strengthening. The circular shape of the cylinders helps keep the block rigid. Struts or webs are also provided to help support the forces exerted by the crankshaft. These cast-in elements are found in both gasoline and diesel engines. However, in a diesel, the reinforcing components tend to be heavier and stronger to withstand the added stresses of diesel operation. This is sometimes referred to as strengthening to the "lower end" of the engine.

Block Variety

At the present time, there is a limited variety of diesel engine types used in private passenger cars. Most of these engines are adapted directly from their gasoline engine counterparts and use similar although strengthened blocks. These blocks include:

1. In-line 4- and 5-cylinder diesel blocks from Mercedes-Benz.
2. Two V–8 blocks from GM.
3. In-line 4-cylinder block from Volkswagen.
4. In-line 4- and 6-cylinder blocks from Nissan Diesel Ltd. used in small International Harvester and Chrysler trucks.
5. In-line 6-cylinder block from Volvo.
6. In-line 4-cylinder block from Peugeot.

CYLINDER HEAD

One of the main jobs performed by the cylinder head is enclosing the top of the combustion chamber (formed in combination with the cylinders and the pistons). The cylinder head also contains or provides space for:

1. Intake and exhaust valves.
2. Intake and exhaust ports.
3. Passages for coolant liquid.
4. In some engines (Mercedes-Benz and Volkswagen), an overhead camshaft and associated valve train components.
5. Openings for fuel injectors.
6. In many engines, openings for glow plugs to improve cold starts.

The cylinder heads of most passenger car diesels closely resemble their gasoline engine counterparts. The biggest difference visually is the absence of the spark plugs and the presence of the machined openings or holes to accommodate the fuel injectors and in some cases, the glow plugs. Another difference, although not so apparent visually, is the extra strengthening that is cast into the heads of the diesel automotive engine. Like the block, the head must be stronger to withstand the additional pressures created by diesel operation. Figure 3–6 shows a typical automotive diesel head.

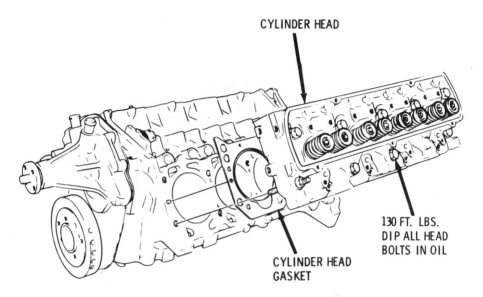

FIGURE 3–5.   **A General Motors cylinder head (*Courtesy of Oldsmobile Division, General Motors Corporation*).**

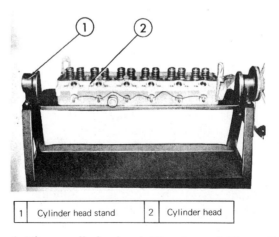

| 1 | Cylinder head stand | 2 | Cylinder head |

FIGURE 3–6.   **A Nissan cylinder head (*Courtesy of Nissan Diesel Motors Ltd./Marubeni America Corporation*).**

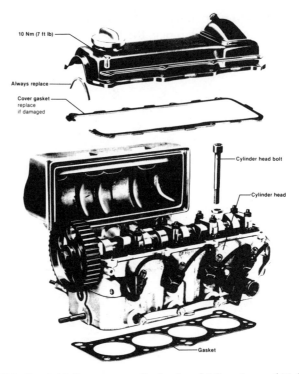

10 Nm (7 ft lb)

Always replace

Cover gasket
replace
if damaged

Cylinder head bolt

Cylinder head

Gasket

FIGURE 3-7. A Volkswagen cylinder head (*Courtesy of Volkswagen*).

Manufacture

The cylinder head of most passenger car diesels (or the cylinder heads of any engine for that matter) are cast in one piece from grey iron or aluminum. After casting, the head is machined to precise shape and tolerances: The upper surface is planed so the head can be attached to the block. The valve seats are machined directly into the head, or otherwise prepared so that specially treated valve seats may be installed. Various openings or holes (for the fuel injectors, glow plugs, valve guides, etc.) are prepared.

After the head is completed, it is attached to the block by large studs or screws installed into the block. The cylinder head is sealed to the machined surfaces of the block by a special heat and pressure resistant gasket. The head is carefully attached by following a strict bolt tightening sequence and by using a torque wrench, applying exactly the torque specified by the manufacturer. If the procedure is not followed exactly, poor performance and early failure of the head and/or gasket will result.

THE CRANKSHAFT

Power from the expanding combustion gases is transferred through the piston, piston pin, and connecting rod to the crankshaft. Each connecting rod is attached at its larger end by a rod bearing to a "throw" on the crankshaft (see Figure 3–8). This offset "throw" receives the thrust of combustion as the piston passes top dead center and as it continues on for 90 to 180 degrees. The crankshaft changes the reciprocating, up and down motion of the piston into a force applied in a circular manner, in other words, torque.

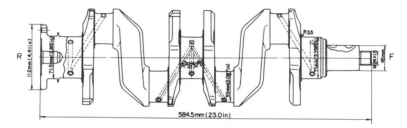

FIGURE 3–8. A Nissan crankshaft (*Courtesy of Nissan Diesel Motors Ltd./Marubeni America Corporation*).

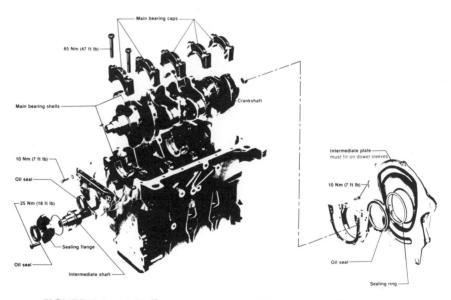

FIGURE 3–9. A Volkswagen crankshaft (*Courtesy of Volkswagen*).

Manufacture

The crankshaft must be rugged enough (both the material and the design) to handle torque throughout its entire length. Either forged or cast crankshafts can be used. Forged shafts are stronger; they are formed by heating a special steel billet and then pressing or hammering it into shape. Cast crankshafts are less expensive and are rarely used in diesel applications. However, improvements in casting technology may result in more crankshafts of this type.

Vibration Damper

Even if the crankshaft is strong and well designed, power impulses occurring first at one end of the shaft and then at the other can sometimes cause trouble. This is especially true if these torsional vibrations occur at the same time or frequency as vibrations caused by other moving parts in the engine. A thumping or bumping noise is heard. And as a result of these harmonic vibrations, parts may fail from metal fatigue.

A device alternately called a vibration damper, torsional vibration damper, or harmonic balancer helps keep these vibrations at tolerable levels. The damper is a small but heavy wheel attached by a special rubber-like material to the pulleys mounted on the front end of the crankshaft. Harmful vibrations occurring in the crankshaft are at least partially absorbed by the mass and inertia of the damper. (See Figures 3–10 and 3–11 for examples of dampers.)

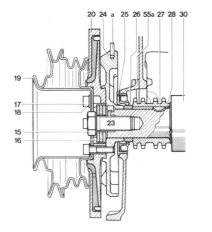

15 Screw, M 18 × 1.5 × 45	25 Radial seal ring
16 Colonical spring washer	26 Spacer ring
17 Screw, M 8 × 30	27 Woodruff key
18 Lock washer B 8	28 Crankshaft sprocket
19 V-Belt pulley	30 Crankshaft
20 Vibration damper	55a Cover
23 Dowel pin, 8 × 8 mm	a Pin for TDC pick up
24 Balancing disc	

FIGURE 3–10. A Mercedes harmonic balancer (*Courtesy of Mercedes-Benz*).

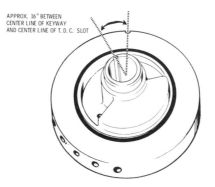

APPROX. 16° BETWEEN
CENTER LINE OF KEYWAY
AND CENTER LINE OF T. D. C. SLOT

**FIGURE 3-11.   A General Motors harmonic balancer (*Courtesy of Oldsmo-
bile Division, General Motors Corporation*).**

Crankshaft Journals and Bearing Surfaces

As noted earlier, the crankshaft is attached to the block by main bearing caps. These half-circular caps clamp over bearings that surround
the crankshaft's main journals. The number of journals depends on the
design of the engine. General Motors V–8s use 5, Mercedes-Benz five-
cylinder engines have 6, and the Volkswagen four-cylinder engine has
5. (See Figures 3–12a, b, rods and mains.)

a

b

**FIGURE 3-12.   Main and connecting rod journals on a crankshaft (*Courtesy
of Gould, Inc.*).**

The circular journal surfaces are machined on a common axis along the centerline of the crankshaft. Often, the end journal will be somewhat larger than the others and contain a flanged thrust surface to handle the end-to-end movement of the crankshaft. This lateral movement must be restricted to several thousandths of an inch. In some engines, the thrust bearing may be located in the middle of the crankshaft; in other engines, it may be at the front.

Besides the main bearing surfaces, journals must be machined for the crank or connecting rod throws. Usually these journals are smaller than the mains.

Lubricating Provisions

Diesel engine crankshafts, like those used in gasoline engines, are drilled with small holes to provide oil flow passages to each main and connecting rod journal. Oil flows under pressure into the space or clearance between the bearing and journal surfaces. When the engine is running, the crankshaft literally "floats" in a cushion of oil so that metal to metal contact between the moving parts is minimized, if not totally eliminated (see Figure 3–13).

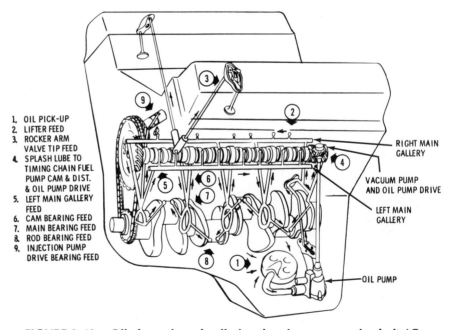

1. OIL PICK-UP
2. LIFTER FEED
3. ROCKER ARM VALVE TIP FEED
4. SPLASH LUBE TO TIMING CHAIN FUEL PUMP CAM & DIST. & OIL PUMP DRIVE
5. LEFT MAIN GALLERY FEED
6. CAM BEARING FEED
7. MAIN BEARING FEED
8. ROD BEARING FEED
9. INJECTION PUMP DRIVE BEARING FEED

RIGHT MAIN GALLERY

VACUUM PUMP AND OIL PUMP DRIVE

LEFT MAIN GALLERY

OIL PUMP

FIGURE 3–13. Oil channels and galleries showing passages in shaft (*Courtesy of Oldsmobile Division, General Motors Corporation*).

Flywheel Flange

A metal flange is machined into the rear of the crankshaft to attach the crankshaft to the flywheel or torque converter. Bolts or screws of high tensile strength join the two elements.

ENGINE BEARINGS

Engine bearings are designed to support and protect the rotating parts of the engine. The engine bearings are also designed to be replaceable. Any surfaces in rotating contact will wear after a period of time. It is better to replace relatively inexpensive bearings than it is to replace an entire crankshaft or camshaft (see Figures 3–14 and 3–15).

FIGURE 3-14. Main bearings precision inserts (*Courtesy of Gould, Inc.*).

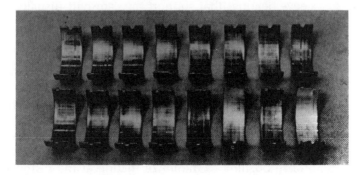

FIGURE 3-15. Rod bearing precision inserts (*Courtesy of Gould, Inc.*).

In varying degrees, all engine bearings have these features in common:

1. **Load carrying capacity.** A bearing must be rugged enough both in design and material to carry the loads imposed on it. In other words, the bearing must not break because of normal mechanical stresses.

2. **Embedability.** It is better for small abrasive particles to sink into the bearing material than to remain on the surface. This prevents the crankshaft or camshaft journal surfaces from being scratched. So the bearing must be thick enough and the material soft enough to allow a reasonable amount of embedability.

3. **Corrosion Resistance.** The bearing must be made of materials that resist the corrosive action of chemicals created in the crankcase (as a result of the combustion process).

4. **Fatigue Resistance.** In addition to resisting immediate mechanical stress, the bearing material must not fail due to the minor bending and twisting present in normal engine operation.

5. **Conformability.** The bearing material must be soft and pliable enough to flow slightly from tight areas to areas with more clearance. In other words, the material must act something like a liquid to partially fill up the bearing space and still leave the desired clearance.

6. **Good Wear Rate.** In addition to being strong, soft, and resistant, the bearing must also have a good wear rate. It must last a reasonable period of time.

Two principal kinds of bearings are used in the diesel engine: the split-ring, which are half-shell bearings that support the crankshaft; and the sleeve bearings that support the camshaft (usually).

Crankshaft Bearings

Each crankshaft bearing (both main and rod) comes in two halves or shells. In a typical main, one shell is installed in a specially machined "upper main saddle" position in the block and the other half is installed in the main bearing cap. The bearings between the large ends of the connecting rods and the crankshaft throws are installed in much the same way. The small end of the connecting rod may have a press fit bronze bushing to minimize friction at the piston pin. The particular bearing used depends upon the design of the assembly.

Following are some particular features of crankshaft bearings:

1. ***Material.*** Crankshaft bearings are usually made of two kinds of materials sandwiched together. The back (e.g., the part that does not come into contact with the rotating journal) is usually made of steel for strength. The inner surface is made from a special alloy, which may include lead, aluminum, bronze, and zinc. This alloy is sometimes called a babbit material. It is softer than steel and designed to provide the features noted in items 2, 5, and 6 above.

2. ***Lubricating Provisions.*** Each bearing shell has holes drilled in it to allow oil to enter the space occupied by the rotating journal. Some bearing surfaces also include an annular groove to control the flow of oil once it has entered the space.

3. ***Installation Features.*** Most precision bearing shells have positioning "tangs" that allow the bearing to be installed and held in the correct position and to prevent the bearing from rotating around the journal once it has been installed.

4. ***Replacement Features.*** Even though bearings are designed to prevent wear on the rotating surfaces, some reduction in journal size does occur over a period of time and under some conditions. Therefore, bearings are available in various sizes to accommodate smaller shaft diameters resulting from resizing or wear.

Camshaft Bearings

Camshafts, depending on the design of the engine, are located in the engine block or the cylinder head. Camshafts located in the block use press fit bearings that are inserted into holes line-reamed through the block webbing. These bearings have the same features as the crankshaft bearings, except that they are formed in one piece. Because they are press fitted, no location tang are required.

Camshafts located in the head use one-piece, press fit bearings, or two-piece clamped bearings, depending on the design of the head. In either case, the bearings have similar features to those mentioned before.

PISTONS, RINGS, AND RODS

As in the case of the other engine parts examined, diesel pistons and connecting rods are stronger than their gasoline engine counterparts to withstand the added pressures and stresses imposed by diesel operation. The top part of diesel pistons may also look different, depending on the design of the combustion chamber. (Various diesel combustion chamber designs, including piston head configurations, are discussed in Chapter 5.)

PISTONS

A piston is actually a cylindrical plug that moves up and down in the cylinder. Most pistons are forged or cast from an aluminum alloy. The top part of the piston is called the head. It contains the circular ring grooves. The lower part of the piston is called the skirt. The piston pins are connected to the piston just below the piston head in the skirt area of the piston. The head is thicker and heavier than the skirt in order to accommodate the ring grooves and withstand the pressures of combustion.

At first glance the piston appears to be a perfect cylinder, however this is rarely the case. If you drew a line connecting the piston pins and another line at right angles to that, the second line would be slightly longer. This elliptical design allows the piston skirt to expand without seizing or sticking in the cylinder as the engine reaches operating temperature. Slots may be cut into the piston just below the piston head. These slots act as a heat dam to prevent the heat of the piston head from over-expanding the skirt. In some designs vertical slots may also be provided to control expansion of the skirt. (See Figures 3–16 through 3–19 for examples of pistons.)

Piston Rings

Since the piston expands and contracts as the engine heats up and cools off, the piston must be somewhat smaller than the cylinder. However, if any significant clearance exists between the cylinder wall and the piston, engine performance suffers. Compression falls off and exhaust gases "blow by" the piston into the crankcase. Therefore, some kind of flexible seal is needed to prevent blow-by and compression loss and at the same time allow the piston to expand and contract. This sealing is provided by the top piston rings. These top compression rings are made of special steel (sometimes chromed or zinc clad) and are installed into the top grooves of the piston head. One or two compression rings are common (see Figures 3–20 and 3–21).

Pistons also have grooves for oil rings. These rings act somewhat like scrapers to remove excess oil that has been purposely splashed or sprayed onto the cylinder for lubrication. Again, depending upon design of the engine, there may be one or two "oil" rings.

FIGURE 3-16. A top view of a piston head (*Courtesy of Oldsmobile Division, General Motors Corporation*).

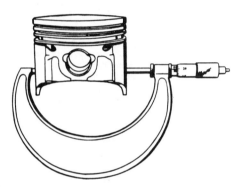

FIGURE 3-17. A side view of a piston (*Courtesy of Oldsmobile Division, General Motors Corporation*).

FIGURE 3-18. A Mercedes piston showing the cooling passages (*Courtesy of Mercedes-Benz*).

FIGURE 3-19. A full view of a Mercedes piston (*Courtesy of Mercedes-Benz*).

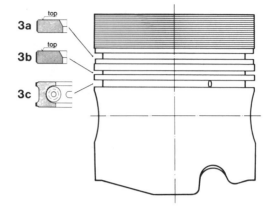

3a   Rectangular ring with inside chamfer (3mm thick)
3b   Rectangular ring with inside chamfer (2 mm thick)
3c   Twin scraper ring with circumferential coil expander (4 mm thick)

FIGURE 3-20.   Mercedes piston rings (*Courtesy of Mercedes-Benz*).

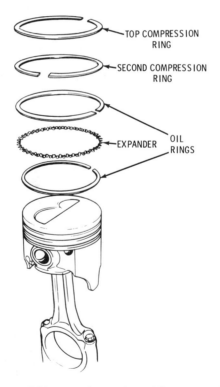

FIGURE 3-21.   General Motors piston rings (*Courtesy of Oldsmobile Division, General Motors Corporation*).

Piston Rods/Connecting Rods

A connecting rod has pivots at both ends to change the reciprocating motion of the piston into crankshaft rotation. The small end of the rod is connected to a relatively large piston pin located in the skirt area of the piston. The movement of the pin with respect to the rod depends on the design of the assembly. If the pin is attached in a fixed manner to the piston skirt, the rod moves about the pin. However, if the rod end is locked to the pin, the pin turns inside a piston pin boss located in the piston skirt area. In either case, the tolerance between the bearing surface and the bearing is usually close.

If the pin moves inside the rod, a bronze bushing is pressed into the small end of the rod. There are three possible ways to handle this kind of arrangement.

1. The pin may be tightly pressed into the rod's small end and held stationary by a lock screw or the tightness of the press fit. In this case, the movement occurs within the machined piston bosses.

2. The piston pin may be pressed or locked with a screw into the piston bosses. In this case, the small end of the rod is provided with a bushing that allows the pin to move within the rod.

3. A full floating pin, allowing rotation both between the piston and the small end rod bushing, may also be used. In that case, it is only necessary to prevent the pin from moving out of the side of the piston, which would score the cylinder wall. Such lateral movement is prevented by a snap ring fitted into special grooves in the piston pin bosses.

The piston rod or connecting rod bearings are subjected to considerable stress and may be one of the first components to fail after long wear. Almost every one, mechanic or not, has heard the phrase, "to throw a rod." The piston rods in diesel engines are particularly stressed. Consequently they are the object of careful design and are usually made of specially forged steel.

Provisions for lubricating the cylinder and rod assembly may be built into the rod. Some engine oiling systems squirt oil through an oil "spit hole" onto the cylinder walls and onto the piston pin. Another design provides a drilled passage through the length of the connecting rod. Oil from the pump flows under pressure through the passages in the crankshaft to the rod throw or journal, through a hole in the bearing shell, and up through the drilled passage in the rod to lubricate the piston pin. Oil thrown off the pin and rod bearings lubricates the cylinder wall. (See Figures 3–22, 3–23, and 3–24 for examples of rods and rings.)

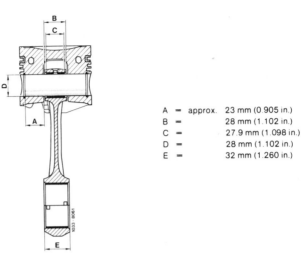

A	=	approx.	23 mm (0.905 in.)
B	=		28 mm (1.102 in.)
C	=		27.9 mm (1.098 in.)
D	=		28 mm (1.102 in.)
E	=		32 mm (1.260 in.)

FIGURE 3-22. A connecting rod and piston cross-section (*Courtesy of Mercedes-Benz*).

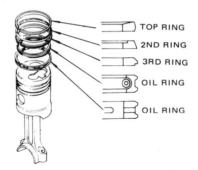

FIGURE 3-23. A connecting rod, a piston, and piston rings (*Courtesy of Nissan Diesel Motors Ltd./Marubeni America Corporation*).

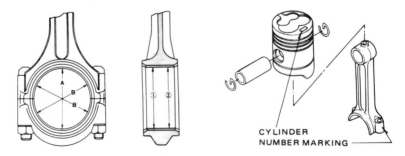

FIGURE 3-24. Construction of a connecting rod (*Courtesy of Nissan Diesel Motors Ltd./Marubeni America Corporation*).

CAMSHAFT, VALVES, AND VALVE TRAIN

These components control the passage of air into the combustion chamber and the flow of exhaust gases out of the chamber. Many European and Japanese manufacturers of automotive diesel engines locate the camshaft, valves, and valve train entirely in the cylinder head. General Motors, the only American diesel manufacturer at this writing, positions the camshaft and valve lifters in the cylinder block; the valves, valve springs, and rocker arms are in the cylinder head. The trend in small gasoline engine development seems to be toward the overhead cam design. It is likely that the diesel engine development will follow this same pattern (see Figures 3–25 and 3–26).

FIGURE 3–25.   A Volkswagen camshaft (*Courtesy of Volkswagen*).

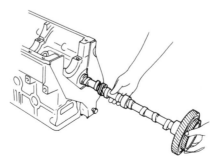

FIGURE 3–26.   A Nissan camshaft (*Courtesy of Nissan Diesel Motors Ltd./Marubeni America Corporation*).

The following paragraphs discuss features common to most camshafts, valves, and valve trains. The accompanying figures illustrate representative examples from the various manufacturers of passenger car diesel engines.

CAMSHAFT

The camshaft is the "brain" of the valve system. The shape and the position of the lobes located along the length of the camshaft "tell" the valves when to open, how much to open, how long to stay open, and when to close. Diesel engine valve overlap (when adjacent intake and exhaust valves are open at the same time) is determined by the position of the lobes on the camshaft.

Manufacture

The camshaft is forged or cast from a special steel alloy into a solid, one-piece unit. Then it is balanced to minimize vibrations and heat treated to control wear. Bearing journal surfaces are machined and polished. The lobes, which are subjected to the most wear, are heat treated to a depth of several thousandths of an inch.

Lobes

There is one lobe or camming surface for each valve. The lobes push against the lifters, which work through the valve train to open the valves (valve closing is accomplished by a valve spring). The lobe design, as previously mentioned, determines the length of time the valves remain open. The spacing of the lobes (around the camshaft axis, not along its length) determines when the valves open and close with respect to one another. Valve overlap occurs when the lobes are positioned so that the intake and exhaust valves are open at the same time (see Figure 3–27).

CAMSHAFT TIMING CHAINS AND BELTS

Since a four-stroke cycle engine requires two crankshaft revolutions per operating cycle, the camshaft is driven at half crankshaft speed. This gives intake and exhaust valve opening and closing for each operating cycle. The camshaft driving mechanism usually depends on the location of the camshaft, whether it is in the cylinder block or in the cylinder head.

FIGURE 3-27. Camshaft lobes (*Courtesy of Volkswagen*).

Cam-in-Block Drives

Many cam-in-block engines, like the GM V–8s, use a timing chain and two sprockets to drive the camshaft. The same sort of mechanism is used in gasoline engines. One sprocket is located on the camshaft. Another sprocket, with half as many teeth, is keyed to the front of the crankshaft, directly behind the harmonic balancer and just inside the timing gear cover. Both sprockets are provided with timing marks. These marks must be in correct alignment after the chain is installed, otherwise, the valves would open at odd and destructive times (see Figures 3–28 and 3–29).

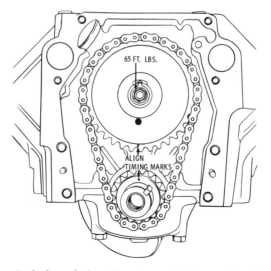

FIGURE 3-28. A timing chain (*Courtesy of Oldsmobile Division, General Motors Corporation*).

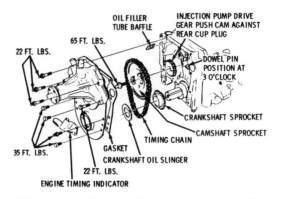

FIGURE 3-29.   Timing sprockets and front engine cover (*Courtesy of Oldsmobile Division, General Motors Corporation*).

Other manufacturers (Nissan, for example) use timing gears instead of chains. However, the ratio between the crankshaft and camshaft gears is still two to one.

Overhead Cam Drives

Some overhead camshaft engines (Mercedes-Benz is one) use a timing
chain to drive the cam; other engines use a reinforced endless rubber
belt. The belt has teeth molded into its inner surface which engage a
driver sprocket on the crankshaft and a driven sprocket on the camshaft. Because the belt is usually quite long, a belt tensioner or ten-
sioning wheel is spring loaded against the outside belt surface. The
tensioner holds the belt taut while at the same time provides a controlled amount of slack for absorbing sudden speed changes and starting loads. (See Figure 3-31 for VW Rabbit belt and Figure 3-32 for Mercedes chain.)

COMPONENTS DRIVEN
OFF THE CAMSHAFT ASSEMBLY

Since the camshaft rotates in precise relationship to engine speed, it is
often used as a driver for other engine parts. Extra lobes, gears, or
sprockets are attached to the shaft of the cam-in-block engines like
General Motors diesel V-8s. The drive belts of the overhead camshaft
engines like the Volkswagen diesel are used to drive or turn an additional sprocket. These and other components are described in the following paragraphs.

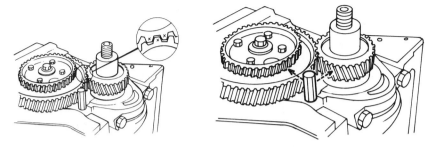

FIGURE 3–30.   Timing gears (*Courtesy of Nissan Diesel Motors Ltd./Marubeni America Corporation*).

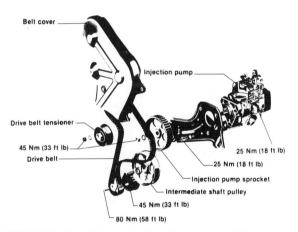

FIGURE 3–31.   Timing belts (*Courtesy of Volkswagen*).

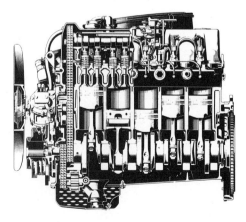

FIGURE 3–32.   Timing chains (*Courtesy of Mercedes-Benz*).

Fuel Injection Pump Drive

GM V–8s use a gear located behind the camshaft sprocket for driving the diesel fuel injection pump. The drive gear is pictured in Figure 3–33.

The four-cylinder Volkswagen diesel uses an extra sprocket driven by the timing belt to operate the fuel injection pump. This arrangement is pictured in Figure 3–34.

Vacuum Pump

Because diesel engines always operate with excess air in the intake manifold, there is no manifold vacuum as such. Devices that are operated by manifold vacuum in gasoline engine applications must be provided with a separate vacuum source in diesels. GM V–8s provide a gear on the camshaft to drive the vacuum pump that supplies the accessory needs of the automobile. The oil pump for the engine lubrication system is also operated by this same gear.

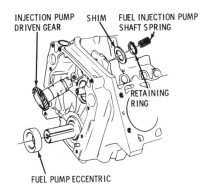

FIGURE 3-33.   Injection pump drive (*Courtesy of Oldsmobile Division, General Motors Corporation*).

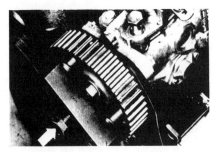

FIGURE 3-34.   Injection pump drive (*Courtesy of Volkswagen*).

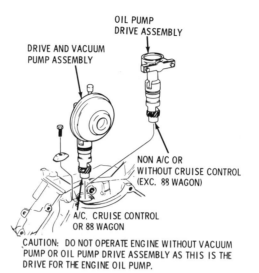

OIL PUMP
DRIVE ASSEMBLY

DRIVE AND VACUUM
PUMP ASSEMBLY

NON A/C OR
WITHOUT CRUISE CONTROL
(EXC. 88 WAGON)

A/C, CRUISE CONTROL
OR 88 WAGON

CAUTION: DO NOT OPERATE ENGINE WITHOUT VACUUM
PUMP OR OIL PUMP DRIVE ASSEMBLY AS THIS IS THE
DRIVE FOR THE ENGINE OIL PUMP.

FIGURE 3-35.   Oil pump and vacuum pump drive (*Courtesy of Oldsmobile Division, General Motors Corporation*).

NOTE

It is interesting to remember that the camshaft of a diesel engine doesn't have the distributor drive gear usually found on the camshaft of gasoline engines (since no electrical ignition system is required). However, the gear provided on the camshaft of GM V-8s to drive the vacuum pump and oil pump is located in the same position formerly occupied by the distributor drive gear.

VALVES AND VALVE TRAINS

Each cylinder in automotive diesel engines has two valves, an intake valve for admitting air into the combustion chamber and an exhaust valve for letting exhaust gases out. As noted before, the opening and closing of the valves is controlled by the camshaft working through a system of rods and linkages collectively known as the valve train. The valves and valve trains of diesel engines closely resemble their gasoline engine counterparts. As in gasoline engines, the biggest difference in diesel valve assemblies is the differences between the valve trains of in-block and overhead camshaft engines (see Figure 3-36).

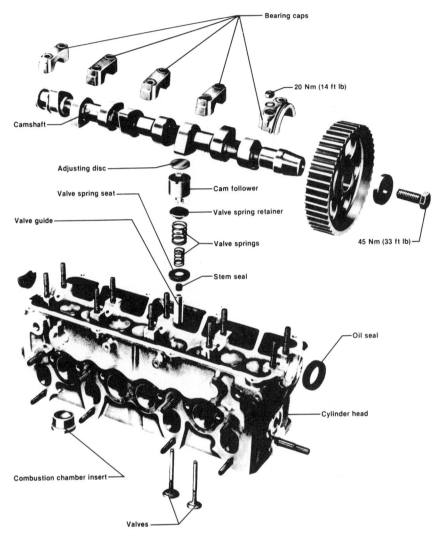

FIGURE 3–36. A Volkswagen cylinder head showing the valve arrangement (*Courtesy of Volkswagen*).

Valves

Intake and exhaust valves are technically known as "poppet valves." They resemble long slim mushrooms that pop open and close. Most automotive poppet valves have certain features in common as shown in Figure 3–37 and described in the following paragraphs.

FIGURE 3-37.   A Mercedes exhaust valve (*Courtesy of Mercedes-Benz*).

Head.   The head or top of the valve closes the port that opens into the combustion chamber. It also is part of the closed combustion chamber.

Margin.   The margin is the area between the head and face of the valve. The margin does not perform any part of the sealing function but strengthens the valve and prevents head distortion. The margin should be not less than 1/32 inch (between the top of the head and the face).

Face.   The face mates with the seat to seal off the combustion chamber. The face and seat are precision ground to an angle of 30 to 45 degrees (usually).

The Seat.   The valve seat may be a hard metal insert pressed into a special machined groove in the head, or it may be a precisely ground integral part of the cylinder head itself. Some manufacturers use the same angle for both the seat and the face. Others allow a one or two degree difference. This difference is called an interference angle and allows the face and seat to make a hairline contact. If the angles are the same, the entire face touches the seat.

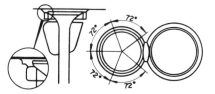

FIGURE 3-38.   Nissan exhaust and intake valve seats (*Courtesy of Nissan Diesel Motors Ltd./Marubeni America Corporation*).

Stem. The stem slides up and down in the valve guide. The guide may be a specially made steel insert pressed into a hole machined in the head, or it may be an integral part of the head itself. In either case, the tolerances are very slight because the amount of lateral stem movement has a direct relationship to the fit between the valve face and the seat.

Retaining Grooves. The valve stem passes up through a coil spring sitting on the head. Two-piece, conical locks or keys (fitting into the retaining grooves and washer-like retainers) connect the valve to the spring. When the valve is opened (via a camshaft lobe acting through the valve train) the valve spring is compressed. After the valve train components stop pushing down on the valve, the spring pushes up on the retainer to close the valve.

Valve Trains

Figures 3-39, 3-40, 3-41, and 3-42 illustrate valve trains of various types of diesel engines. (See also Figure 3-7.) As you can see, most are fairly simple mechanisms whose operation is easily deciphered. Valve opening is accomplished by rods, rocker arms, etc. Valve closing takes place when the valve springs that were compressed during valve opening are allowed to relax. However, the following noted GM feature does bear particular mention.

GM Hydraulic Valve Lifters. (See Figure 3-42.) This type of valve lifter maintains zero clearance or lash between the push rods and rocker arms, thus eliminating the periodic adjustment necessary in mechanically operated systems. Following is a simplified explanation of how hydraulic valve lifters work: Inside the lifter body is a plunger that rests on a disc valve and a plunger return spring. Oil goes from the side of the lifter body into the plunger and then on up into the hollow push rod. When a cam lobe pushes up on the lifter, all the oil passageways are closed and the lifter, plunger, and push rod rise together as a unit. As the unit goes up, a certain amount of oil leaks from the base of the lifter. By the time the valve has fully opened, enough leakage has taken place so the valve spring does not encounter any resistance when it pushes the valve closed. Then, when the lifter goes back down the cam lobe, the plunger return spring pushes the plunger up, which lets more oil into the lifter and returns the valve train to zero clearance. The cycle is repeated every time the valve is opened.

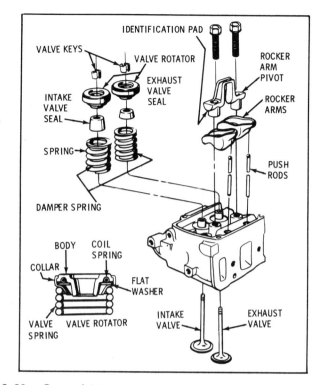

FIGURE 3-39.   General Motors valves and parts, including rotators (*Courtesy of Oldsmobile Division, General Motors Corporation*).

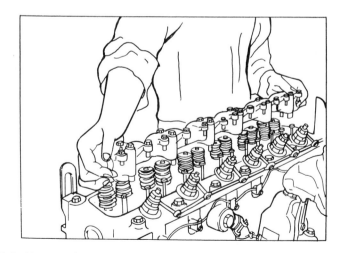

FIGURE 3-40.   A Nissan rocker arm assembly (*Courtesy of Nissan Diesel Motors Ltd./Marubeni America Corporation*).

FIGURE 3–41.   A mechanical valve lifter (tappet) and push rod (*Courtesy of Nissan Diesel Motors Ltd./Marubeni America Corporation*).

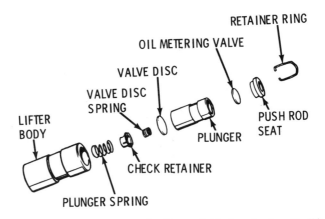

FIGURE 3–42.   The components of a General Motors hydraulic lifter (*Courtesy of Oldsmobile Division, General Motors Corporation*).

CHAPTER 4

Performance Measurements and Terminology

GENERAL

You have probably been exposed to the terms work, energy, power, friction, etc. You probably also have some understanding of what TDC, BDC, stroke, bore, compression ratio, etc., mean. In this chapter, we will review these performance-related terms and concepts, expanding on the knowledge you already possess to create a framework of precise definitions that will serve as building blocks for understanding more complex aspects of diesel operation.

PISTON POSITION TERMS

As you have already seen in this book, abbreviations are used to describe the position of the piston in the cylinder: whether it is at the top of its travel, the bottom, or somewhere in between.

TDC (Top Dead Center)

TDC refers to the apex or highest point the piston travels in the cylinder. Although TDC could be applied to any piston when it is at the top of its travel, the term is most often used to describe the position of the piston in the number one cylinder. The position of the number one piston serves as a reference point for other engine operations: adjusting the fuel injection pump, timing the valve train, etc. Usually the particular reference point is specified by the manufacturer. In other words, a particular adjustment will be made at so many degrees before top dead center (BTDC), after top dead center (ATDC), or at zero degrees TDC.

The location of the number one piston with respect to TDC is determined by comparing an indicator on the harmonic balancer (discussed in the previous chapter) with a fixed reference point. The particular form of the indicator varies from manufacturer to manufacturer. Some companies machine a narrow groove across the outer circumference of the harmonic balancer (see Figure 4-1). Others press a steel bead into the balancer. In either case, when the mark and the pointer (or other nearby mark) line up, it lets you know that the number one piston is at top dead center. Additional graduations are often added to determine degrees BTDC or degrees ATDC.

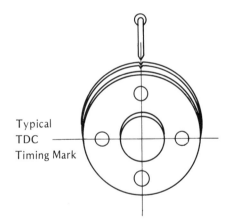

Typical
TDC
Timing Mark

FIGURE 4-1.   Harmonic balancer, timing mark, and pointer.

Note
Some manufacturers locate the timing mark on the face of the flywheel. The mark is viewed through a hole in the bell housing and compared with the fixed reference pointer to find TDC.

BDC (Bottom Dead Center)

BDC is the extreme opposite of TDC; it refers to the lowest possible piston position in the cylinder (see Figure 4-2). Usually there is no indicator showing BDC and it is rarely used as a reference for other engine operations.

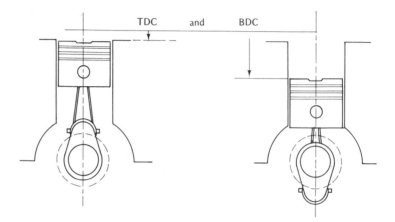

FIGURE 4-2.   Top dead center (TDC) vs. Bottom dead center (BDC).

ENGINE SIZE/DISPLACEMENT TERMS

The size of an engine is often given in terms of its displacement: the combined working area of all its cylinders. Displacement is either expressed in inches (e.g., 305 cubic inches displacement) or liters (e.g., 2.5 liters). The following paragraphs show how displacement is determined.

Stroke

Stroke refers to the movement of a piston from TDC to BDC, or how far it travels from one end of the cylinder to the other. This figure is given either in inches (e.g., 3.150″) or in metric units (e.g., 80.00 mm).

Crankshaft Throw

The stroke is directly related to the crankshaft throw—the offset of the crankshaft connecting rod journal with respect to the main bearing journal. For instance, if the offset is 2″ (if the centerline of the main bearing journal is 2″ from the centerline of the connecting rod bearing journal), then the stroke will be twice that distance, or 4″. The throw and corresponding stroke are the same for every piston of a given engine (see Figure 4–3).

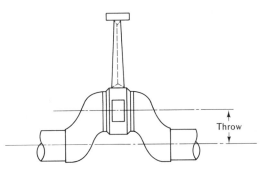

FIGURE 4–3.   Crankshaft throw.

Bore

The bore is the diameter of the cylinder and is the same for all the cylinders in a given engine. Like the stroke, the bore is stated in inches (e.g., 3.012″) or in metric units (e.g., 76.50 mm). Both the bore and the stroke are used to find the size of the engine and to make power calculations. The bore size also serves as a reference point when measuring for cylinder wear and when resizing a cylinder when an engine is being rebuilt (see Figure 4–4).

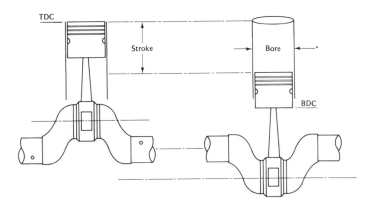

FIGURE 4–4.   Bore and stroke.

Displacement

This is the volume displaced by one piston as it moves from TDC to BDC. In other words, it is the volume of an imaginary cylinder whose height is the same as the stroke of a piston and whose diameter is the same as the bore of the cylinder. The formula for finding the displacement of a cylinder is the same as the formula for finding the volume of a cylinder (see Figure 4–5).

$$\text{Volume} = \pi \times R^2 \times \text{Stroke}$$
$$\text{or}$$
$$\text{Volume} = 3.14 \times \tfrac{1}{2}\ \text{bore}^2 \times \text{Stroke}$$

The volume of the entire engine is obtained by multiplying the above figure times the number of cylinders in the engine. These figures may be in inches (cubic inches of displacement, e.g., CID) or in metric units (cubic centimeters, e.g., CCs or liters).

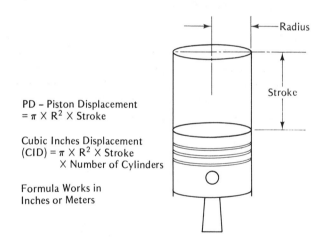

PD – Piston Displacement
$= \pi \times R^2 \times$ Stroke

Cubic Inches Displacement
(CID) $= \pi \times R^2 \times$ Stroke
        $\times$ Number of Cylinders

Formula Works in
Inches or Meters

Radius

Stroke

FIGURE 4–5.   Formula for calculating cubic inch displacement (CID).

Volume

Sometimes the displacement of an engine is confused with the total volume of the engine. The total volume is the displacement PLUS the volume remaining in the combustion chamber when the piston is at TDC. This is sometimes called the clearance volume because it represents the clearance between the top of the piston and the head (see Figure 4–6).

Total Volume = Clearance Volume +
Displacement Volume

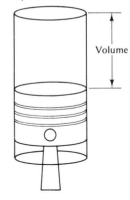

FIGURE 4-6.   Cylinder volume.

Compression Ratio

As noted in Figure 4-7, the compression ratio is a comparison between two volumes. It is the ratio or relationship between the total volume in a cylinder and the clearance volume of the cylinder. The greater the ratio, the more air is squeezed in the compression stroke. As we have seen in previous chapters, diesels operate at high compression ratios, meaning the air is squeezed a great deal.

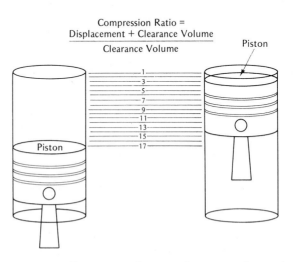

FIGURE 4-7.   Clearance volume and compression ratio.

Volumetric Efficiency

This is another comparison between two volumes—in this case, between the volume of air that theoretically could enter the cylinder and the volume of air that actually does enter the cylinder. It is a measure of how efficiently the engine breathes. The ratio is based on a standard air temperature of 60 °F at sea level pressure (14.7 pounds per square inch). The volumetric efficiency of most well-designed diesel engines is on the order of 85% (for non-boosted, naturally aspirated diesels).

BASIC PHYSICAL CONCEPTS

Before we see how power is measured, we need to briefly examine the basic concepts of work, force, energy, power, etc.

Force

Force is an effort exerted against a resistance. When you push against a wall you exert force.

Work

Work is a combination of force and movement. If you push against a wall and the wall doesn't move, you are not doing any work, no matter how hard you push. However, if the wall moves, then you are doing work. The typical unit of measuring work is the foot-pound. If a ten-pound object is moved 10 feet, 100 foot-pounds of work have taken place.

Energy

Energy is the ability or potential to do work. Diesel fuel possesses chemical energy, which is converted by the engine into useful work. Electrical energy from the battery is converted into work by the cranking motor. Mechanical energy is transmitted from the piston to the crankshaft and on to the flywheel, the driveline, and finally the driving wheels. Although energy can neither be created or destroyed (except in thermonuclear devices), it can be degraded into lower and lower forms until it can no longer be put to practical use. Heat energy radiated from the radiator or exhaust is an example of low-grade, wasted energy.

Torque

Torque is force applied in a circular or twisting manner. When you turn or twist on a screwdriver handle you are exerting torque. The same idea applies when the crankshaft is rotated by the piston and connecting rods. Torque is measured in pounds-feet (not foot-pounds). Figure 4–8 illustrates torque and how it is measured.

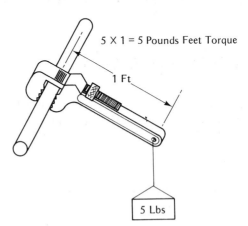

5 X 1 = 5 Pounds Feet Torque

1 Ft

5 Lbs

FIGURE 4–8.   Torque means turning or twisting force.

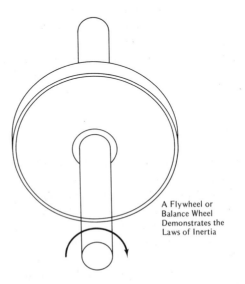

A Flywheel or
Balance Wheel
Demonstrates the
Laws of Inertia

FIGURE 4–9.   The laws of inertia control the operation of a flywheel.

Inertia

If you have ever pushed a car by hand, you probably noticed that it was more difficult to get the car moving than it was to keep it moving. In day-to-day driving, you have probably also noted that once your vehicle is moving it displays a stubborn tendency to go in the direction it is headed. It has to be forced to go around corners, forced to slow down, forced to speed up. This is inertia, the resistance of an object to changes in speed or direction. The engine flywheel makes use of inertia (see Figure 4–9) by resisting changes in speed to dampen out uneven firing impulses and to help balance the engine. The harmonic balancer at the front of the crankshaft operates in the same fashion to minimize torsional vibrations along the crankshaft.

FRICTION

Friction is either friendly or unfriendly depending on where it is found, and on what kind of friction it is.

Dry Friction

This is just what the name says, resistance to sliding movement between two dry surfaces. The friction between the brake shoe and the drum, or between the brake pad and rotor, is an example of dry friction. Without this kind of friction, you could not stop your car. Without dry friction between the tires and the road, a car could not move forward or be kept on the road. In these cases, dry friction is a friendly phenomenon. However, dry friction isn't so friendly when it takes place between two moving or rotating parts inside the engine—for instance, between a journal and a bearing surface.

Greasy Friction

As much as possible, dry friction is eliminated between rotating or sliding engine parts. You might think that dry friction could be eliminated by supplying a layer of grease or heavy lubricant between the bodies. Up to a point this is true. However, at high speeds, thick greases separate and let the engine parts come into contact. As a result, friction and wear between the parts is not reduced appreciably.

Viscous Friction

This is the friction or resistance to movement that occurs between the molecules of a liquid (see Figure 4–10). If the proper lubricant is introduced between two, high-speed, sliding or rotating bodies, the bodies themselves will not touch and friction will be limited to resistance to motion between layers of the lubricant. That is basically how oil protects the engine parts. (Diesel lubrication will be discussed in more detail in Chapter 11.)

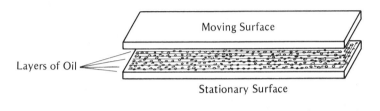

Moving Surface

Layers of Oil

Stationary Surface

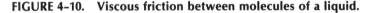

Viscous Friction

FIGURE 4–10.   Viscous friction between molecules of a liquid.

POWER

Everybody has an idea of what power is. We say a person is powerful (physically, mentally, emotionally, etc.). We describe a car or machine as being powerful. In each case, we are describing the ability of someone or something to perform in certain ways. The technical definition of power is much the same except that we add one ingredient that might not have been present in our informal definition—time. In a technical sense, power is the ability to do a certain amount of work in a certain period of time. There are a number of ways to measure power and various units used to express how much power a device can produce. The following paragraphs describe some ways used to measure the power produced by internal combustion engines.

Horsepower

This is the most common way to express the power produced by an engine. The term originated when draft horses were still in common use. Engineers and scientists rather arbitrarily decided that one large horse could lift 33,000 pounds one foot in one minute. So an engine that could do the same amount of work in the same time was said to produce one horsepower.

The problem is how do you measure horsepower? How do you find out how much work an engine can do in a certain period of time?

It is not practical to match an engine with a team of horses. Some other method must be used. The next paragraphs note some of the methods that have been used.

S.A.E. Horsepower

One of the earliest ways to describe the horsepower of automobile engines was developed by an organization called the Society of Automotive Engineers. They didn't actually measure how much work an engine does over a period of time, but instead used this formula to indirectly compare the output of engines:

$$\text{S.A.E. H.P.} = \frac{\text{Bore Diameter}^2 \times \text{Number of Cylinders}}{2.5}$$

This formula is simple; anyone can use it to find how much power an engine can produce. Unfortunately, though, the formula doesn't take into account such factors as compression ratio, engine speed, length of stroke, etc. As a result, S.A.E. horsepower is rarely used except for legal or licensing purposes.

Brake Horsepower (BHP)

Brake horsepower is a direct measurement of the usable power produced by an engine. It is the most common way to determine the output of an engine.

The term brake, as related to horsepower, originated with a device called the *prony brake*. It was a forerunner of the modern dynamometer and consisted of a variable braking device connected by a lever to a weighing scale. The arm resting on the scale applied resistance to a wheel turning inside the brake. The engine being tested was connected to the wheel. The ability of the engine to turn the wheel at certain revolutions per minute (RPMs) and under certain loads was measured and the resulting figures inserted into a formula to determine its brake horsepower. Figure 4–11 pictures a prony brake and gives the formula for determining brake horsepower.

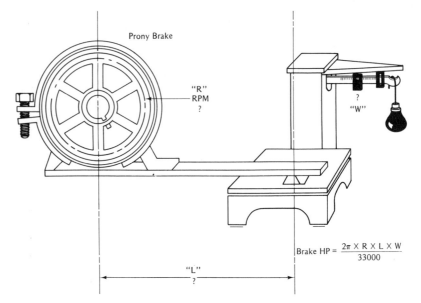

Prony Brake

"R"
RPM
?

?
"W"

Brake HP = $\dfrac{2\pi \times R \times L \times W}{33000}$

"L"
?

FIGURE 4-11. The Prony Brake was a device for measuring horsepower.

Today, brake horsepower can be checked at the end of the crankshaft using an engine dynamometer, or at the drive wheels using a chassis dynamometer. Some modern dynamometers are absorption units and work on the principle of the prony brake. Other dynamometers are actually electrical generators connected to the engine. The brake horsepower of the engine is determined by measuring the amount of electrical power produced (746 watts = 1 hp) and converting that figure to brake horsepower.

SPECIALIZED HORSEPOWER RATINGS

In addition to these common ways of measuring and stating horsepower, engineers use special measuring techniques to describe and compare various aspects of engine operation.

Indicated Horsepower

Indicated horsepower is a relative measurement of the power produced by burning fuel in an engine's cylinders. First, the mean effective pressure produced during the four strokes of operation is determined by electronic testing devices in a laboratory. Then, that reading is inserted into the following formula:

$$IHP = \frac{P \times L \times A \times N \times K}{33,000}$$

Where P = Mean effective pressure in pounds per square inch (PSI)
           L = Length of the stroke in feet
           A = Cylinder area in square inches
           N = Power strokes per minute $\left(\frac{RPM}{2}\right)$
           K = Number of cylinders

Frictional Horsepower

This is a measurement of the power lost due to friction—in other words to the resistance of engine parts in sliding contact. The test is performed by driving an engine by an electrical motor and measuring the resistance offered. Tests have shown that friction increases with speed and that as much as 70 to 80% of friction losses occur in the cylinder area.

TEMPERATURE/HEAT

Temperature and heat are important concepts in engine operation. Temperature measurements are constantly being made by various devices, and automatic adjustments are made as a result of these measurements. Heat relates to the basic operation of the engine itself, since the engine is a device for converting chemical energy into heat energy, which is then converted into mechanical energy.

Atomic Structure of Matter

Before talking about the related concepts of heat and temperature, it is important to remember that all substances—you, this book, all materials in the universe—are made up of tiny particles called atoms. Atoms, in turn, are made up of even smaller particles called electrons, protons, and neutrons. Heat and temperature are measurements of the activity within the atoms of matter.

Temperature

Atoms and their component parts are always in motion. Temperature is a measure of that motion. High temperature means more atomic activity; lower temperatures mean less activity.

Heat

Heat is a measurement not only of how much atoms vibrate, but of how many atoms are in vibration. In other words, heat is a measurement of mass as well as activity. Look at it this way: a large body of water at 33 °F contains more heat than a small kettle of boiling water at 212 °F. There is more total activity (atomic) in the large body of water than in the small kettle.

Temperature Measurement Scales

There are two common scales for measuring temperature: fahrenheit and celsius. Both are based on the freezing and the boiling points of water at sea level atmospheric pressure (14.7 pounds per square inch or PSI). In the United States, the fahrenheit scale of 32 (freezing) to 212 degrees (boiling) is most commonly used. However, the trend is toward the metric or celsius method, which sets the freezing point of water at 0 degrees and the boiling point at 100 degrees.

Heat Measurement Scales

Since heat is a measurement of quantity as well as activity, it must take into account mass as well as temperature. There are two common ways to measure heat: the BTU (British Thermal Unit) and the calorie. The BTU is the heat required to raise the temperature of one pound of water one degree fahrenheit (see Figure 4-12). A calorie is the heat required to raise the temperature of one gram of water one degree celsius (actually from 3.5 to 4.5 °C). This means that if you eat a doughnut containing 100 calories, you have consumed enough food energy to raise one gram of water 100 degrees—from freezing to boiling.

BTU = Heat Required to Raise One Pound of Water One Degree Fahrenheit

FIGURE 4-12.   BTUs are units of heat measurement.

TEMPERATURE MEASUREMENT DEVICES

The most common temperature measurement device is a bulb thermometer like the one shown in Figure 4-13. It consists of a glass bulb containing liquid mercury or colored water and a graduated column connected to the bulb. The bulb is placed near the substance whose temperature is being measured (outdoors in the air, under the tongue, in water, etc.). Atomic activity in the substance being checked is transferred to the liquid in the bulb causing the liquid to expand or contract. As the liquid changes volume, it goes up or down in the column. The height of the liquid is compared to the graduated scale and read as temperature.

Although automotive thermometers are hardly ever of the glass bulb type, they still respond to atomic activity to indicate temperatures.

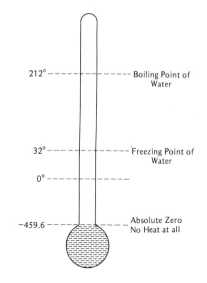

FIGURE 4-13.   A glass bulb thermometer.

Thermocouple

This kind of temperature sensing device consists of two dissimilar wires welded together at one end and connected to a sensitive meter at the opposite end. (See Figure 4-14.) Heat applied to the ends joined together creates atomic activity that the meter registers as electrical current flow. By properly calibrating the meter, electrical readings can be converted to temperature readings. Heat gauges in automobiles often use thermocouples to convert heat energy into electrical energy, which is then measured and displayed as temperature.

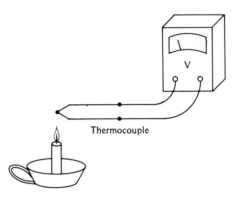

FIGURE 4–14.   A diagram of "thermocouple".

Bi-Metal Thermometers

The bi-metal coil or spring is another kind of temperature sensing device that makes use of dissimilar metals joined together. In this case, strips of dissimilar metals are bonded, sandwich-like, into layers. When the bi-metal thermometer is exposed to heat, the two metals absorb atomic energy and expand. However, because their atomic structures are different, the two metals expand at different rates. As a result, the strip tends to curl, or, if fashioned into a coil shape, to wind up or unwind (see Figure 4–15).

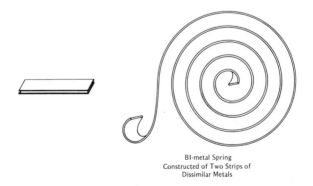

BI-metal Spring
Constructed of Two Strips of
Dissimilar Metals

FIGURE 4–15.   Bi-metal springs react to changes in temperature.

Bi-metal type thermometers are usually used to signal temperature changes to some type of electrical or mechanical device. The bi-metal strip is attached to a set of electrical contacts or a small lever or an indicator pointer calibrated in degrees.

EXPANSION

As noted above, when materials absorb atomic activity from other, more active sources, expansion occurs. The fractional amount a metal expands for a one degree rise in temperature is called its coefficient of expansion. Expansion rates of all materials used are taken into account by engineers when they design an engine. Temperature regulation devices are built into the engine to sense and control temperatures so that expansion stays within designed limits.

CHAPTER 5

Combustion Chamber Design

GENERAL

The combustion chamber is the heart of the diesel engine (or any internal combustion engine). Its design affects thermal efficiency, emission control, operating smoothness, noise, etc. In one way or another, all engine operations are designed to come together in the combustion chamber—combining air and fuel, then burning the mixture to release the energy stored in the fuel.

There are two common diesel combustion chamber designs: the precombustion/turbulence design and the open chamber design. Both satisfy the basic diesel requirement of creating then controlling high velocity air and fuel flow. However, each design employs a somewhat different approach and each has its own advantages and disadvantages.

At the present time, all automotive manufacturers use the precombustion chamber design. We'll examine both precombustion/turbulence and open combustion chambers. Then we'll look at some specific combustion chambers made by various automotive diesel manufacturers.

PRECOMBUSTION CHAMBERS

A precombustion design actually consists of two chambers in one. The prechamber itself is an egg-shaped cavity, part of which is formed separately then press fitted into a machined opening in the head (one prechamber per cylinder). The prechamber contains the tip of the glow plug and fuel nozzle assembly and is located adjacent to the intake and exhaust valves. The other 70% to 75% of the combustion chamber clearance volume is located over the piston in a conventional manner and is connected to the prechamber by passageways (see Figure 5-1).

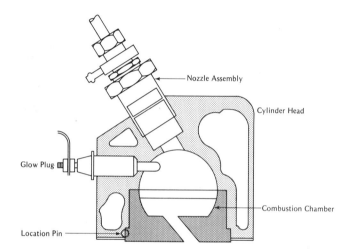

Nozzle Assembly

Cylinder Head

Glow Plug

Combustion Chamber

Location Pin

FIGURE 5-1. Precombustion chamber, diesel combustion chamber design.

As the name suggests, combustion originates in the prechamber. High pressure air is forced into the chamber during the compression stroke. Then, at just the right moment before TDC, the fuel injectors open up. As the first stage of ignition takes place and pressure builds up in the prechamber, a jet of flame and hot fuel vapor squirt into the main chamber. The flame and hot vapor mix with the fuel and air already there to complete the combustion process.

TURBULENCE COMBUSTION CHAMBER

The turbulence combustion chamber performs in almost the same way as the precombustion chamber. These are the principal differences (see Figure 5-2):

1. The turbulence chamber is larger, occupying as much as 75% to 80% of the clearance volume.
2. The entire turbulence chamber is usually cast integrally with the head.
3. Usually the fuel injector nozzle projects directly into the turbulence chamber; it is recessed in most prechamber designs.

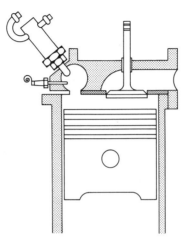

FIGURE 5-2. Turbulence chamber, diesel combustion chamber design.

OPEN CHAMBER DESIGNS

The open-type, combustion chamber more closely resembles the design of conventional gasoline engine combustion chambers. The entire clearance volume is located over the piston; there are no chambers or cavities in the head. However, there are two cone-shaped depressions in the top of the piston (see Figure 5-3). As the piston moves up in the compression stroke, these cavities cause the air to swirl about in different directions. By the time the centrally located injector starts spraying fuel into the piston depressions, the air is moving fast enough to carry hot vapor and flame throughout the combustion chamber.

ADVANTAGES AND DISADVANTAGES

Both major designs have their strong and weak points. Generally, an open chamber design provides higher thermal efficiency and power. However, in the areas that relate most closely to passenger car requirements, the prechamber/turbulence chamber design is superior as noted in the following.

1. Emission control is easier with the prechamber/turbulence design.
2. Because the ignition delay period (refer back to Chapter 2) is longer in the prechamber/turbulence type design, the engine tends to run smoother and is quieter.
3. The prechamber/turbulence design is less sensitive (in terms of timing and fuel injector) than the open chamber design.

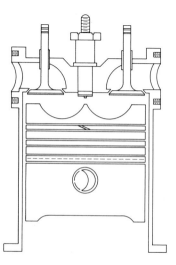

FIGURE 5-3.   Open chamber, diesel combustion chamber design.

EXAMPLES OF SPECIFIC
COMBUSTION CHAMBER DESIGNS

The following paragraphs and pictures describe some specific diesel combustion chambers in use at the time of this writing. Cylinder head information where appropriate is also included.

Peugeot Cylinder Head and Combustion Chamber Design

The prechamber configuration used by Peugeot is called a swirl design because the arrangement of prechamber components promotes a swirling air fuel flow. Special heat-treated swirl chambers are pressed into precisely machined depressions in the head adjacent to the valves. The swirl chambers—one for each cylinder—are not allowed to protrude below the flat lower surface of the head more than 0 to .03 mm. The opening from the swirl chamber angles out into the main chamber in the direction of the valves.

The head of the four-cylinder Peugeot automotive diesel is a cast, cam-in-head unit. The valves sit side-by-side longitudinally along the block. When seated, the valve heads are recessed 0.75 to 1.15 mm within the head. See Figure 5–4 for details.

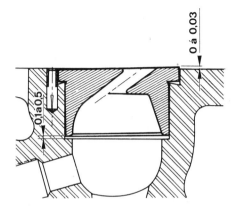

FIGURE 5–4.   Peugeot swirl chamber (a prechamber design).

Mercedes-Benz 5-Cylinder Diesel Prechamber Design

Mercedes (along with Audi) produces one of the few 5-cylinder engines on the market and like the Peugeot, the Mercedes diesel is also a cam-in-head engine with the usual features common to that type of design. However, as noted in Figure 5–5, the prechamber is shaped somewhat differently. The tip of the Mercedes prechamber protrudes below the surface of the head into depressions shaped in the top of the piston. Note that Figure 5–5 shows two openings leading from the prechamber tip into the main chamber. The larger opening (on the left) goes to the larger dished-out portion of the piston so that proportionally more flame and hot vapor are squirted into that part of the clearance volume. Besides the 5-cylinder diesel, Mercedes also produces a 4-cylinder diesel automotive engine.

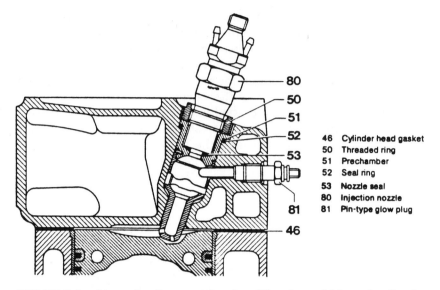

46	Cylinder head gasket
50	Threaded ring
51	Prechamber
52	Seal ring
53	Nozzle seal
80	Injection nozzle
81	Pin-type glow plug

FIGURE 5-5.   Mercedes-Benz prechamber (*Courtesy of Mercedes-Benz*).

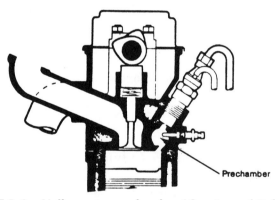

FIGURE 5-6.   Volkswagen prechamber (*Courtesy of Volkswagen*).

Volkswagen 4-Cylinder Automotive Diesel Head and Combustion Chamber Design

The 4-cylinder VW diesel engine also employs an overhead cam design with intake and exhaust valves sitting side-by-side lengthwise along the centerline, almost exactly above the cylinders. The cylinder head, with the valves installed and seated, is almost uniformly flat with slight recesses at the prechamber access openings. Machined into the top of each piston is a clover leaf design to allow clearance for valve opening. At TDC, when the full 23.5 to 1 compression ratio is achieved, the pistons may protrude as much as .040 mm above the top of the cylinder block.

The VW prechambers are made from special steel alloys. The prechambers are pressed into openings in the lower cylinder head and are replaceable when the engine is overhauled (see Figure 5–6 for VW diesel).

General Motors V–8 Diesel Head and Combustion Chamber Design

The cylinder head configuration of the GM automotive diesel is very similar to the design of the typical American gasoline V–8. There is a camshaft in the block, hydraulic lifters, push rods, rocker arms, etc. The valves sit side-by-side over each cylinder, and when seated, the valve heads protrude below the head about the distance of the valve head margin. Otherwise, the cylinder head is flat, except for recessed openings at the prechamber discharge points.

The prechambers are recessed into the head surface, one above each piston. The prechambers are replaceable and are installed with a mallet, flush with the head surface (±.003). The tops of the piston are notched (see Figure 5–7 for GM prechambers) to provide clearance at TDC when the compression ratio has reached 22.5 to 1.

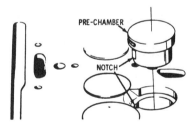

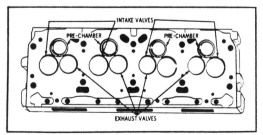

FIGURE 5–7.   General Motors 5.7 liter diesel prechamber (*Courtesy of General Motors Corporation*).

Nissan SD 22 and SD 33 Head and Combustion Chamber Design

Nissan makes 4- and 6-cylinder diesel engines for automotive and small truck applications. Both types employ a cam-in-block, prechamber design. The prechamber, as noted in Figure 5–8, is similar in appearance to the Peugeot swirl chamber. The Nissan swirl-type prechamber sits flush with the head surface. The valve heads, when seated, are not allowed to protrude more than .029 mm into the combustion chamber.

The design of the Nissan combustion chamber, like that of the other engines, is completed by the configuration of the piston head. It contains a double leaf, relief design to provide valve clearance. The compression ratio is 22.0 to 1.

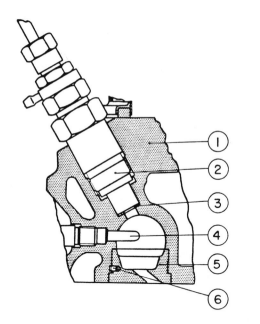

1	Cylinder head
2	Nozzle assembly
3	Gasket
4	Glow plug
5	Combustion chamber
6	Knock pin

FIGURE 5-8. Nissan prechamber (*Courtesy of Nissan Diesel Motors Ltd./Marubeni America Corporation*).

CHAPTER 6

General Engine Disassembly Procedures

GENERAL

As we've seen in previous chapters, diesel and gasoline automotive engines are quite similar in appearance—the principal differences being some parts not shared by both and the fact that even those components that do look alike are usually heavier and stronger when used in diesels. The same kind of similarity holds true for general engine disassembly procedures. The basic steps are nearly the same for both types of engines, the principal differences occurring at the detailed disassembly level.

This chapter examines those generally accepted procedures common to most types of automotive engine disassembly. And though a number of illustrations are provided to show disassembly details for some specific diesel engines, reference should be made to the manufacturer's shop manual before actually attempting to take an engine apart. It will give the detailed instructions that cannot be included in a general book like this.

CLEANING THE ENGINE

The first step in any engine disassembly operation is to clean the engine. The road grime, dirt, and abrasives that coat the engine cause enough trouble on the outside. However, if this material gets inside the engine when parts are removed and the interior exposed, serious damage can result. It's like preparing to do surgery on the human body. An incision would not be made without first cleaning the surrounding skin.

An engine can be cleaned before or after it is removed from the chassis. However, most rebuilders prefer to do it after the engine is out because the parts are more accessible and a more thorough cleaning job can be performed.

Whether the engine is cleaned before or after it is removed from the chassis, it is often necessary to do some minor disassembly work before the actual cleaning begins. Any parts (such as alternators, fuel injection pumps, air cleaners, etc.) that might be damaged by the cleaning method should be removed from the engine and the openings plugged. If the parts aren't removed, they should be carefully covered with waterproof material and any potential leakage points taped up.

After the engine has cooled to the surrounding temperature (recommended by GM) any one or a combination of several cleaning methods can be used: steam, high pressure water and detergent, or the time honored stiff brush and solvent method. The latter approach, although more troublesome, does allow time for detailed scrutiny of parts and assemblies for defects or wear.

REMOVING THE ENGINE

Taking the engine out of the engine compartment is dangerous and the manufacturer's directions must be followed exactly. Consequently, we won't go over any detailed removal procedures, other than to note the following general guidelines common to all types of automobiles.

1. Use professional lifting equipment and make sure it is in good repair. A block and tackle hung from the limb of a shade tree is romantic, but not very safe.
2. Make sure the work area is clean and uncluttered. If the engine slips, you need a clear escape route.
3. Make sure you have all the special tools recommended by the manufacturer.

4. Protect the fenders or any other parts of the car that might be scratched or otherwise damaged.

5. Before lifting the engine, check to be sure all wires, cables, coolant hoses, fuel lines, etc., have been disconnected and secured out of the way.

6. If the automobile itself has to be raised, make sure it is properly supported on adequate stands. Chock the wheels.

MOUNTING THE ENGINE ON A WORKSTAND

After removing the engine from the engine compartment, most professional rebuilders use a workstand to support the engine. This gives better access both for cleaning and later on for disassembly.

Various kinds of workstands are available. Some are designed by automotive manufacturers for specific engines. Others, made by aftermarket suppliers, are adaptable to a number of different kinds of engines. A good stand should not offer any obstructions to stripping the engine down to the block. The best stand should have a gear and crank assembly that lets the worker rotate the engine to the best position for disassembly (and later on for reassembly). Figure 6–1 shows a Nissan diesel engine mounted on a rotary work stand.

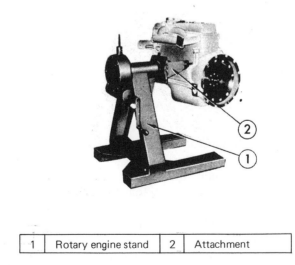

| 1 | Rotary engine stand | 2 | Attachment |

FIGURE 6–1. A Nissan engine on a rotary stand. (*Courtesy of Nissan Diesel Motors Ltd./Marubeni America Corporation*).

Before mounting an engine on a work stand, it is often necessary to remove some engine components and attach a mounting bracket to the engine. In the case of the Nissan shown in Figure 6–1, removing the alternator and its bracket and the oil gallery plug makes available three holes for attaching an engine mounting adaptor. Figure 6–2 shows the mounting adaptor installed on an engine.

FIGURE 6–2.   A Nissan engine with adapter attachment in position (*Courtesy of Nissan Diesel Motors Ltd./Marubeni America Corporation*).

REMOVING THE ENGINE COMPONENTS

The remaining paragraphs in this chapter discuss the usual sequence of parts removal and some commonly accepted considerations for removing these components.

Removing the Oil Pan

After the engine has been mounted on a stand and cleaned, most manufacturers recommend that the oil pan be removed first. However, before the oil pan is taken off, the oil should be drained and properly disposed of (not thrown out back to seep into the ground water). After the oil has been drained, screw the drain plug back in so it isn't lost. Now the engine can be rotated on the workstand so that the screws holding the oil pan in place are easy to reach (see Figure 6–3).

As in each of the following steps, the parts, screws, bolts, etc., that are removed from the engine should be set aside immediately after being taken off. Parts should be put in separate locations or containers and in some cases should be labeled for reassembly.

Removing the Manifolds

At this point, the engine is usually rotated back into an upright position. The air cleaner is removed (if it wasn't already taken off) and the intake and exhaust manifolds are detached from the cylinder head assembly. Figures 6–4 and 6–5 show the procedure for removing these components from a Nissan diesel engine. Naturally, the technique will vary somewhat from engine to engine. For instance, if the engine were equipped with a turbocharger, like the Mercedes 5-cylinder, then the charger would have to be removed first (see Figure 6–6).

Cranking Motor Removal

As noted in Figure 6–7, the cranking motor is usually removed next.

Cylinder Head Removal

The cylinder head is now detached in most disassembly procedures. However, before it can be taken off, there are usually some other parts that must be removed first. The particular items depend primarily on whether the engine is a cam-in-block or a cam-in-head unit.

Cam-in-Block. Remove the valve cover, fuel injectors, injector pump, glow plugs, rocker arms and/or rocker arm shaft, and the push rods.

Cam-in-Head. Remove the timing chain/belt cover, the valve cover, the camshaft, fuel injectors, injector pump, and glow plugs (or at least the glow plug harness).

Nut/Screw Removal. Now, the stud nuts or head screws that hold the head in place can be undone. They should be removed in the reverse order of the tightening sequence (found in the shop manual for the particular engine being disassembled).

After the head has been carefully lifted off the studs, it should be examined for damage or excessive wear. Then all the parts should be lightly oiled to prevent them from rusting before the engine is reassembled.

Figure 6–8 shows details of GM diesel head removal. Figures 6–9a and b picture the procedure for removing the head from a Nissan diesel engine.

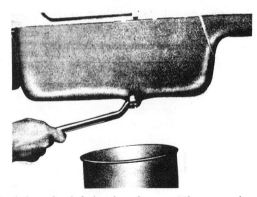

FIGURE 6-3. Draining the lubrication from a Nissan engine before disassembly (*Courtesy of Nissan Diesel Motors Ltd./Marubeni America Corporation*).

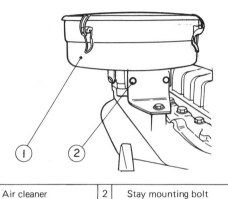

1	Air cleaner	2	Stay mounting bolt

FIGURE 6-4. A Nissan air cleaner assembly (*Courtesy of Nissan Diesel Motors Ltd./Marubeni America Corporation*).

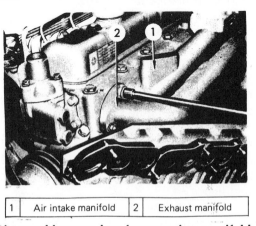

1	Air intake manifold	2	Exhaust manifold

FIGURE 6-5. Disassembly procedure for removing manifolds from a Nissan diesel engine (*Courtesy of Nissan Diesel Motors Ltd./Marubeni America Corporation*).

FIGURE 6–6.   Removal of a Mercedes turbocharger (*Courtesy of Mercedes-Benz*)

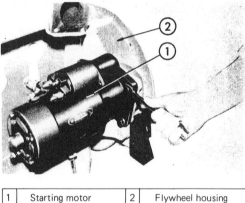

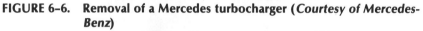

1	Starting motor	2	Flywheel housing

FIGURE 6–7.   Removal of a Nissan starter motor (*Courtesy of Nissan Diesel Motors Ltd./Marubeni America Corporation*).

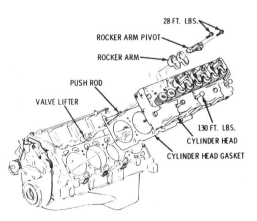

FIGURE 6–8.   Details of head and rocker arm removal in a General Motors 350CID diesel engine (*Courtesy of General Motors Corporation*).

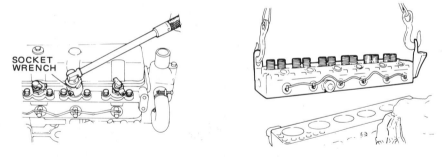

FIGURE 6-9.   Removal of a Nissan cylinder head (*Courtesy of Nissan Diesel Motors Ltd./Marubeni America Corporation*).

Camshaft Removal

If the engine is a cam-in-head unit, the camshaft will already have been removed at this point. However, if cam is located in the block, it is usually taken out now. First, the front pulley or balancer is detached from the engine. Then the timing gear or chain cover is removed, the chain or gear inspected and removed, and the locating plate bolts taken off. Finally, after making sure the valve lifters and push rods have been removed, the camshaft is extracted from the engine. After being examined for damage or wear, it is lightly oiled and set aside. Figure 6-10 shows camshaft removal in a Nissan diesel engine.

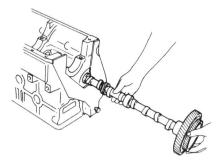

FIGURE 6-10.   Removal of a Nissan camshaft (*Courtesy of Nissan Diesel Motors Ltd./Marubeni America Corporation*).

Oil Pump Removal

To remove the oil pump, first carefully detach the pickup screen. All that remains then in most cases is to loosen some screws and lift the pump out. After being carefully cleaned and examined, the oil pump is set aside.

Connecting Rod and Piston Removal

Following are the steps commonly followed in this stage of disassembly.

1. The cylinder walls are examined above the highest point of ring travel. If a ridge exists, it is removed with a ridge reamer.

2. Connecting rods and rod caps must be reinstalled exactly in their former positions. Therefore they are checked for identification or markings. If none exist (and they don't in some diesel as well as some gasoline engines), the cylinder number must be inscribed on each rod and cap. This is done with a center punch and hammer, or some kind of engraving or etching device.

3. Now the screws or nuts holding the rod caps can be removed and set aside—in the order they were removed. Next, the rod cap is carefully removed.

4. At this point it is usually recommended that plastic sleeves (actually, sections of special hose available at most parts houses) be installed over the rod bolts. This protects the surface of the crankshaft as the rods and pistons are pushed out the top of the cylinders. Figure 6–11 shows these sleeves installed on the rod bolts of a GM diesel.

5. Now the piston and rod assemblies are lifted up through the top of the engine, care being taken not to damage anything.

6. It is a good idea before going any further to connect the appropriate rods and rod caps back together.

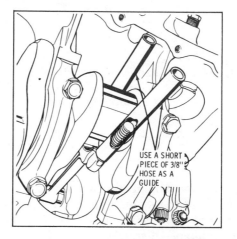

FIGURE 6–11. A connecting rod bolt guide and journal protector sleeve (*Courtesy of General Motors Corporation*).

Main Bearing and Cap Removal

Owing to the nature of the line-reaming process, main bearing caps will fit in only one location, even though they appear identical. So, before being removed, the main bearing caps must be examined for marks identifying their location. If none are present, identifying marks must be punched or scribed.

Crankshaft Removal

The last item removed is the crankshaft. It is lifted out with care, both to avoid damage to the crankshaft and to the person handling it. The crankshaft is quite heavy and is sometimes removed using a rope or fiber sling and hoist. Figure 6-12 illustrates crankshaft removal from a Nissan diesel. After the crankshaft has been set aside in a protected location, the upper main, half-shell bearings can be lifted out.

FIGURE 6-12.   Removing the crankshaft from the engine block with a sling (*Courtesy of Nissan Diesel Motors Ltd./Marubeni America Corporation*).

SUMMARY

Remember, engine disassembly can be dangerous. Care should be taken and the manufacturer's directions followed exactly. Do not attempt to perform an engine disassembly without the appropriate tools and reference literature.

CHAPTER 7

Air Intake System

GENERAL

The air intake system is responsible for delivering clean air in adequate amounts to the cylinders. This chapter examines the operation of diesel air intake systems and notes some specific examples of systems employed by various manufacturers of automotive diesels.

COMPARISON WITH GASOLINE ENGINES

Gasoline and diesel air intake systems are similar with one important exception. Gasoline engines usually contain an air valve or throttle. The throttle opening determines the volume of air and fuel going to the cylinders, which, in turn controls the speed of the engine.

Most diesel engines, on the other hand, do not contain an air valve. The speed of the engine is regulated by the quantity of fuel injected. Air enters freely at all times, the amount taken in, or "aspirated" being determined by the RPM and the displacement of the engine. As a result, there is no manifold vacuum—unlike gasoline engines where the pistons "pull" a vacuum in the intake manifold during closed or part throttle operation. This is the reason that automotive diesels need a separate vacuum pump to operate various auxiliary devices.

COMPONENTS

The two main intake components in a non-boosted diesel engine (boosted engines are discussed later in the chapter) are the air cleaner and the intake manifold. These components serve the same purpose in diesels as they do in gasoline engines.

Air Cleaner

The air cleaner strains particles from the air that might damage the engine and quiets the noise of air rushing into the engine. Most manufacturers use the familiar paper-element-type replaceable filter housed in a metal case. Figure 7–1 shows the paper filter and air cleaner used by GM on its automotive diesel engines. Only Peugeot uses the oil-bath-type oil cleaner (and then, in only one of their models). It is pictured in Figure 7–2. This type of filtering system is effective, but requires careful maintenance.

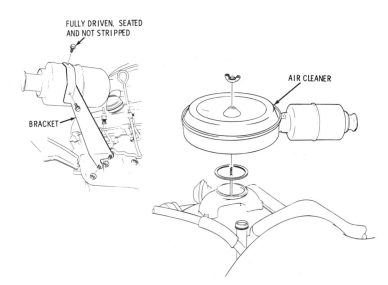

FIGURE 7–1. An air cleaner housing and how it is attached (*Courtesy of Oldsmobile Division, General Motors Corporation*).

FIGURE 7–2. A completely assembled oil bath air cleaner (*Courtesy of Peugeot*).

Intake Manifold

The intake manifold is a system of passageways that lead from the air cleaner at one end to the intake openings in the cylinder head at the other end. An intake manifold may be a simple tube-shaped casting attached to the side of an in-line engine. Or it may be a complex casting sitting between the heads of a V–8 engine.

Figures 7–3 and 7–4 show the intake manifold used on GM V–8 diesels. Note in Figure 7–4 that the intake manifold has two openings on top that are connected to an air crossover assembly, which in turn is connected to the air cleaner.

Figure 7–5 pictures the intake manifold used on a six-cylinder, in-line Nissan diesel engine. Since the intake ports for cylinders one and two, three and four, and five and six use adjacent cylinder head openings, the intake manifold branches into three main channels, one leading to each pair of openings.

Figure 7–6 shows the intake manifold used on a four-cylinder, in-line Nissan diesel engine. This manifold is also grouped into branches leading to adjacent cylinder head openings with the air cleaner located on top and more or less centered between the cylinders.

In all of these engines, the intake manifold is connected to the head by cap screws or by studs and nuts. The seal to the head surface is obtained by using special heat-resistant gaskets that are both strong enough and pliable enough to make sure no air leaks occur. This insures that all the air entering the engine comes through the air cleaner.

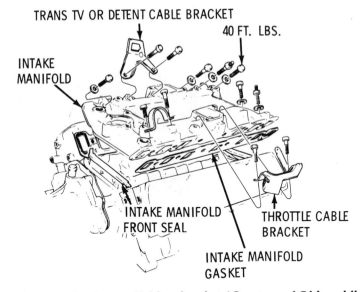

FIGURE 7-3. An intake manifold and gasket (*Courtesy of Oldsmobile Division, General Motors Corporation*).

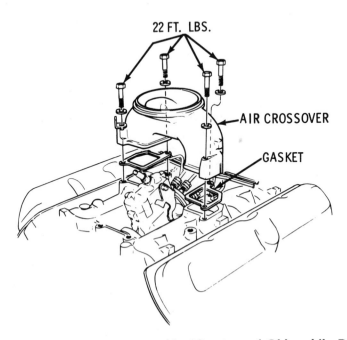

FIGURE 7-4. Air cross-over assembly (*Courtesy of Oldsmobile Division, General Motors Corporation*).

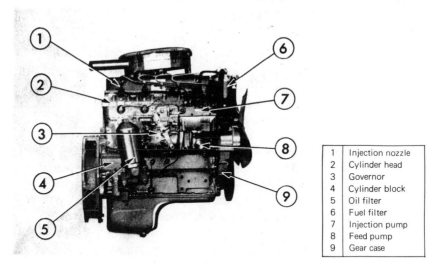

1	Injection nozzle
2	Cylinder head
3	Governor
4	Cylinder block
5	Oil filter
6	Fuel filter
7	Injection pump
8	Feed pump
9	Gear case

FIGURE 7-5. An intake manifold and air cleaner for a 6-cylinder Nissan diesel (*Courtesy of Nissan Diesel Motors Ltd./Marubeni America Corporation*).

1	Air cleaner
2	Intake manifold
3	Thermostat housing
4	Exhaust manifold
5	Cooling fan
6	Generator
7	Fan belt
8	Starting motor
9	Flywheel housing
10	Oil pan

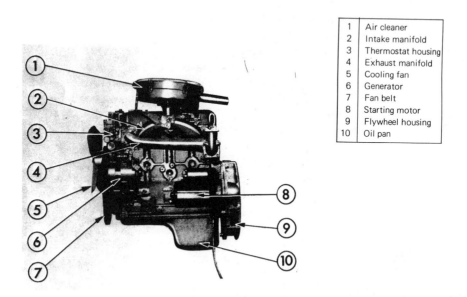

FIGURE 7-6. An intake manifold and air cleaner for a 4-cylinder Nissan diesel (*Courtesy of Nissan Diesel Motors Ltd./Marubeni America Corporation*).

NATURALLY ASPIRATED VERSUS
BOOSTED DIESEL ENGINES

Aside from leaks, there are two ways to bring air into an engine. One way is to let the air enter naturally at normal atmospheric pressure. This is called natural aspiration. The other way is to force the air into the engine under pressure by means of a mechanical device. This is called supercharging or boosting.

Many truck and heavy equipment diesel manufacturers use boosted engines. However, at this time, only Mercedes and Nissan produce boosted automotive diesels, although other companies might go to "blown" diesels as competition becomes keener.

The remaining paragraphs in this chapter briefly examine the history of boosters and some types commonly used.

HISTORY OF SUPERCHARGERS

The idea of increasing the pressure of air going into the engine dates back to the early days of internal combustion engines. However, the greatest push to develop superchargers came in the Second World War when aircraft companies had to find a way to increase the operating altitude of bombers and fighter planes. At 10,000 feet, air pressure is only about 7.5 PSI (compared to 14.7 PSI at sea level). Naturally aspirated engines trying to operate at such altitudes will suffer "shortness of breath" due to the lack of oxygen. So, to increase pressure and thereby improve breathing, a centrifugal, impeller-type charger was designed to operate between the carburetor and the intake manifold. By forcing more air into the engine, the supercharger made up for the oxygen deficiency. After the war, pressure boosters became fairly common in large truck diesels. In fact, many two-cycle diesels require a blower of some kind in order to run at all.

TYPES OF BLOWERS

Although various types of supercharging devices are used to increase the performance of internal combustion engines, three types are most likely to be found in diesel application: centrifugal/impeller blowers, rootes blowers, and turbochargers.

Centrifugal/Impeller Blowers

This type of unit typically consists of a compressor-type fan connected to a drive belt, which in turn is connected to the crankshaft. As the compressor is rotated by the drive belt, it forces air under pressure into the diesel intake manifold.

Rootes Blower

This type of blower also forces air under pressure into the intake manifold. However, instead of using a fan to compress the air, it uses two, three-lobe rotors turning in a special housing. As the lobes mesh and unmesh, they force air through the housing, acting somewhat like a two-gear-type oil pump in an engine lubricating system.

Turbochargers

A turbocharger is a kind of centrifugal blower. However, instead of using a belt to drive the air compressor wheel, the energy contained in the exhaust gas is employed. The turbocharger is located between the exhaust manifold and the exhaust pipe. As exhaust gas passes through the turbocharger it rotates a turbine wheel. This wheel is connected on a common shaft with a separate compressor wheel which rotates as the turbine rotates. The compressor wheel forces air under pressure into the intake manifold. The amount of pressure or "boost" varies with the speed of the engine and can reach dangerous levels. So, it is usually limited in most engines by safety devices built into the turbocharger.

Turbochargers are the most popular type of boosters used on truck diesels. Figure 7–7 shows the exhaust and intake air flow in the Mercedes system. Figure 7–8 pictures the unit as it appears when mounted on the engine.

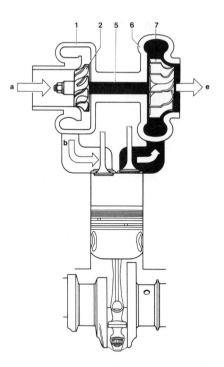

1   Compressor housing
2   Compressor wheel
5   Shaft
6   Turbine housing
7   Turbine wheel
a   Compressor intake (fresh air)
b   Compressor discharge (compressed air)
c   Exhaust gas to bypass canal
d   Exhaust gas to turbine wheel
e   Exhaust gas discharge (to exhaust pipe)

FIGURE 7-7.   Air and exhaust gas flow in a turbocharger (*Courtesy of Mercedes-Benz*).

FIGURE 7-8.   A turbocharger in place on a Mercedes diesel (*Courtesy of Mercedes-Benz*).

CHAPTER 8

Diesel Exhaust Systems

GENERAL

Compared to gasoline engine exhaust systems, the exhaust systems found in diesel cars and light trucks are very simple. This situation might change if diesels are subjected to similar anti-pollution legislation. However, at the moment, diesel exhaust systems are characterized by NOT having these components commonly associated with many gasoline engines:

1. Catalytic reactor mufflers.
2. Pumps and associated devices for injecting air into the exhaust stream.
3. Pipes running to the air intake system to preheat air entering the engine.

Instead, diesel exhaust systems are designed solely to perform the basic tasks of:

1. Collecting gases from the exhaust ports.

2. Channeling the gases into a common passage.
3. Passing the gases to a muffler, which breaks up the exhaust pressure waves and thereby quietens the engine.
4. Delivering these gases to the tailpipe where they are vented into the atmosphere.

The following paragraphs briefly review the basic exhaust system components.

EXHAUST MANIFOLD

A typical exhaust manifold (see Figure 8-1) is a collection of passageways formed into a one-piece, cast-iron assembly. Exhaust manifolds are bolted to cylinder heads so that the exhaust ports open into individual passages in the manifold. These passages lead to a larger, central passageway connected to the exhaust pipe or a cross-over tube.

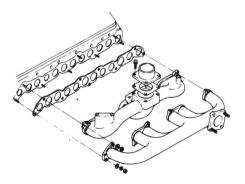

FIGURE 8-1. An exhaust manifold (*Courtesy of Nissan Diesel Motors Ltd./Marubeni America Corporation*).

In-line engines generally have one exhaust manifold. Vee-type engines always have two, one for each bank of cylinders.

Variations in exhaust manifolds are usually due to differences in valve and head design. Figure 8-2 shows the exhaust manifold and exhaust port openings of a typical six-cylinder, in-line engine. Note that the center ports are so close together that the exhaust manifold passageways give the appearance of a double tube. Figure 8-3 shows a typical exhaust manifold and port arrangement for a V-8 engine.

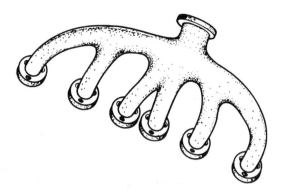

FIGURE 8-2   Typical 6-cylinder exhaust manifold arrangement.

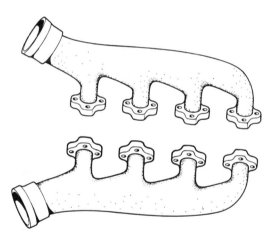

FIGURE 8-3.   Typical 8-cylinder diesel exhaust manifold configuration.

EXHAUST MANIFOLD CONNECTIONS

Some exhaust manifolds are attached to the cylinder heads by cap screws; other use studs and nuts. Some use special, heat resistant gaskets where the machined surface of the manifold mates with the machined surface of the head; others use no gasket at all.

Figure 8-4 shows the cap screw method of manifold attachment used by GM on their V-8 diesels. Figure 8-5 pictures the studs used by Peugeot to attach exhaust manifolds to their in-line, four-cylinder diesel.

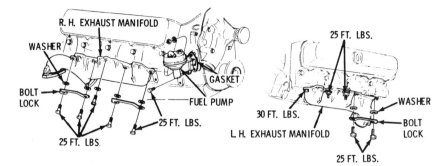

FIGURE 8-4. Manifold attachment bolts and locks (*Courtesy of Oldsmobile Division, General Motors Corporation*).

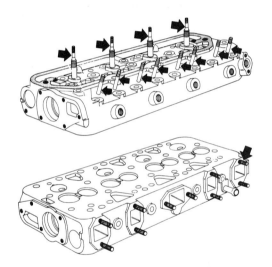

FIGURE 8-5. Typical 4-cylinder head showing the arrangement of the valve parts (*Courtesy of Peugeot*).

CROSS-OVER TUBES

Unless a vee-type engine has dual exhaust with independent exhaust manifolds (or "headers"), a cross-over pipe is required to connect the two manifolds. In a GM V-8 diesel, the cross-over pipe passes under the engine. It leads from the single flanged opening on the left exhaust manifold to a similar opening on the right manifold. The exhaust gases from the two sides of the engine combine and then pass out of a second flanged opening on the right manifold into the exhaust pipe.

EXHAUST PIPE, MUFFLER, TAIL PIPE

After passing into the exhaust pipe, gases go through the muffler and on into the tail pipe (or pipes). The arrangement and design of these components vary from manufacturer to manufacturer.

TURBOCHARGED DIESEL EXHAUST SYSTEM

At the time of this writing, the most "exotic" automotive diesel exhaust system is found on turbocharged Mercedes-Benz and Nissan diesels. As noted in a previous chapter, a turbine is built into the exhaust system of this car. The turbine captures the force of gases rushing out of the engine, using them to compress air going into the engine, thereby improving engine performance. However, downstream of the turbine, in the direction of the exhaust pipe, muffler and tail pipe, the exhaust is essentially conventional.

Figure 8–6 shows a bottom view of the Mercedes exhaust manifold and turbocharger. The pipe pictured in the drawing is an oil return line to the oil pan. Oil flows first from a connection on the oil filter base through a high pressure line into passages in the turbocharger housing. After providing lubrication and cooling, the oil flows back to the sump by the line shown.

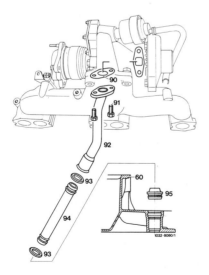

60	Upper oil pan housing
90	Gasket
91	(2) Screw, M 8 × 20
92	Upper oil return line
93	O-ring
94	Lower oil return line
95	Special profile seal ring

FIGURE 8–6. A Mercedes diesel exhaust manifold showing the turbocharger and oil return pipe (*Courtesy of Mercedes-Benz*).

CHAPTER 9

Diesel Fuel and Injection Systems

INTRODUCTION

Diesel fuel and fuel injection systems are key elements in diesel operation. They also relate to some of the most significant differences between diesel and gasoline engines. Consequently, this is one of the most important chapters in the book.

DIESEL FUELS

When buying gasoline, the selection is either based on octane ratings (high test or regular) or whether the gasoline is leaded or unleaded. Diesel fuel selection, on the other hand, is based on different criteria.

In the broadest sense, any fuel that will burn in any of the variety of diesel engines can be termed diesel fuel. Ordinary kerosene will work in many engines. Heavy oils will work in others.

The fuels used in most truck and automotive diesels generally fall into two categories: number one and number two diesel fuels. Number one fuel is lighter, like kerosene, and generally found in colder climates. Number two fuel is thicker, like heating oil, and is used in milder climates. Before discussing other differences between these two fuels, it will be helpful to briefly cover the terms viscosity and cetane rating.

Viscosity

Viscosity refers to how easily a liquid flows. Molasses is a high viscosity liquid; it flows poorly. Water is a low viscosity liquid; it flows easily. Not only does viscosity depend on the properties of a particular liquid, it also depends on temperature. Diesel fuel, for instance, doesn't flow as well at lower temperatures as it does at higher temperatures.

The flow rate of diesel fuels, which are relatively low viscosity liquids, is important for several reasons. If the viscosity is too low (if the liquid is too thin), the fuel loses some of its ability to lubricate the fuel injectors. If the viscosity is too high (if the liquid is too thick), the spray pattern at the injector nozzle tips will be adversely affected. This condition can be particularly pronounced at lower temperatures and results in poor burning. Of course, the fuel also wouldn't burn properly if the viscosity were too low, just as the fuel injectors wouldn't be lubricated properly if the viscosity were too high (instead of too low as mentioned first).

Cetane Ratings

A diesel fuel's cetane number refers to how easily it ignites. The higher the number, the easier ignition. If the cetane rating is too low, the fuel resists ignition and accumulates in the combustion chamber. Then, when the unburned fuel does ignite, late in the firing cycle, severe knocking is likely to result. Higher cetane fuels, on the other hand, tend to ignite instantly, providing smooth burning and even power flow.

Differences Between Number One and Number Two Diesel Fuels

As a practical matter, many private diesel operators will never have to select between number one and number two diesel fuels. That is because both may not be available where they live. Number one fuel is uncommon in many warmer regions. Even in the North, drivers may not find number one fuel, only 2W, which means number two winterized—in other words, number two with some number one added.

This lack of selection may be beneficial because, in some respects, number two is the better fuel. It has a higher BTU or energy content per gallon than number one. It also has a lower sulphur content, which means it tends to burn more cleanly. The principal advantage of number one is its ability to flow more easily in extremely cold weather operation. It also has a higher cetane rating, 44 compared to 40 for number two.

Other Diesel Fuel Facts

1. At the time of this writing, the quality control of diesel fuel doesn't seem as high as gasoline. Some diesel operators report mileage and performance differences from tank to tank even when the kind of fuel used and the driving habits remain the same.

2. Operators also report mileage differences that are probably not due to the fuel but to the fill-up technique used. Diesel fuel tends to foam more than gasoline and will cause automatic pump nozzles to shut off before the tank is filled. To completely fill a tank with diesel fuel, operators often must wait several times during the course of the operation to let the foam settle before the fill-up can continue.

3. Many large fleet operators of diesel vehicles use blends of diesel fuels especially prepared for their own operation.

FUEL FILTERS

Fuel filters are especially important in diesel operation. The parts of the injector pumps and nozzles are so carefully machined to such slight tolerances that the slightest amount of foreign matter will cause abnormal wear or outright failure.

Several types of fuel filters are in common use. Figure 9–1 shows a Nissan fuel filtering unit. It includes a special paper filter as well as an overflow valve. The overflow valve performs several functions. It helps maintain constant fuel pressure in the supply system. It also serves as an air bleed to reduce foaming. The fuel discharged from the overflow valve is sent back to the fuel tank by a fuel return line.

Figure 9–2 shows a Peugeot filter. It uses a replacement type cartridge filter with a recommended change interval of 12,500 miles. Most automotive and small truck diesels are equipped with filters similar to these two, requiring regular replacement of filter elements.

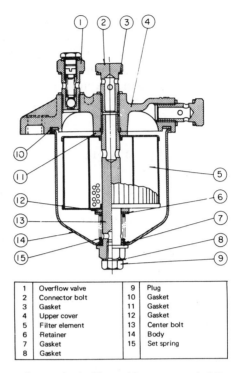

1	Overflow valve	9	Plug
2	Connector bolt	10	Gasket
3	Gasket	11	Gasket
4	Upper cover	12	Gasket
5	Filter element	13	Center bolt
6	Retainer	14	Body
7	Gasket	15	Set spring
8	Gasket		

FIGURE 9-1. A Nissan fuel filter (*Courtesy of Nissan Diesel Motors Ltd./Marubeni America Corporation*).

FIGURE 9-2.   A Peugeot fuel filter (*Courtesy of Peugeot*).

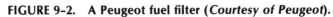

FUEL INJECTION SYSTEM, AN OVERVIEW

The basic functions performed by any diesel fuel injection system include:

1. Storing the fuel.
2. Filtering the fuel.
3. Pumping the fuel to the cylinders in the proper amount at the proper time (in the latter case, performing somewhat the same role as the distributor in a gasoline engine).
4. Injecting the fuel into the cylinders.
5. Advancing and retarding the fuel injection.
6. Governing the speed of the engine to make sure it stays in a maximum/minimum range and does not run away.
7. Returning unused fuel back to the fuel tank.

Diesel fuel injection systems share some of these functions with the injection systems occasionally found in gasoline engines. However, it is important to remember that diesel engines, unlike gasoline engines, REQUIRE fuel injection. For high compression diesel engines to work properly, the fuel must be introduced into the combustion chamber at just the proper instant and in just the right amount. The moment of ignition must be precisely controlled, which is not possible with a carburetor-based system that mixes the fuel and air before it reaches the combustion chamber.

DIESEL FUEL INJECTION SYSTEMS

There are two main types of injection systems used in small truck and automotive diesels. One type, known as the multi-plunger pump, is produced by a number of manufacturers under license from the Robert Bosch Company. It is characterized by having one pump and delivery valve for each cylinder in the engine.

The other type of injection unit is called a distributor pump. It is used by GM, VW, and others and is characterized by having one pump to supply fuel to outlets leading to each of the engine cylinders. In certain respects it is akin to a gasoline ignition distributor that routes high voltage surges from one ignition coil to a number of cylinders.

THE MULTI-PLUNGER, BOSCH SYSTEM

Figure 9–3 pictures the main components of a typical Bosch type system. Summarizing the flow of fuel through the system:

1. Fuel is drawn from the fuel tank by the feed pump.
2. The feed pump then forces fuel through the filter.
3. Excess fuel is sent from the filter back to the fuel tank.
4. The remainder of the fuel goes to the fuel injection pump.
5. Operating via a camshaft running the length of the assembly, the fuel injector pump sends fuel under pressure out of the nozzle pipes.
6. The nozzle pipes carry the fuel to injector nozzles located in each of the cylinders.
7. Excess fuel goes to the fuel tank.
8. The remainder of the fuel is injected into the heated air in the combustion chamber. As noted in a previous chapter, injection typically begins just before the piston reaches TDC and continues for a few degrees after TDC.

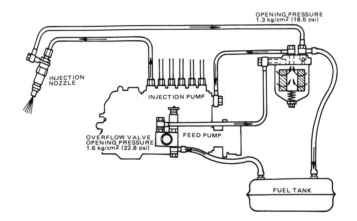

FIGURE 9–3. A fuel injection system (*Courtesy of Nissan Diesel Motors Ltd./Marubeni America Corporation*).

INJECTION PUMP

The heart of the injection system is the injection pump assembly. Figures 9–4, 9–5, and 9–6 picture a Bosch-type pump. The main components of this and similar pumps include:

1. The injection pump camshaft running the length of the lower end of the assembly.
2. An advance/retard timing device (not shown) connected to the end of the camshaft.
3. Roller tappets riding on each of the camshaft lobes.
4. Plunger-type fuel pumping elements riding on top of each tappet.
5. A control rack running the length of the upper end of the pump. This device engages pinions (gear teeth) on each plunger pump to control the amount of fuel injected and thereby determines engine speed.
6. A governor connected to the control rack, acting as an intermediary mechanism between the control rack and accelerator pedal.
7. A fuel gallery or supply chamber located in the top part of the pump and connected to ports in each fuel plunger pump.
8. A spring-loaded delivery valve at the top of each plunger pump.
9. High pressure fuel lines connecting each delivery valve with an injector nozzle located in each of the cylinders.

The following paragraphs describe these components in more detail.

Fuel Injection Camshaft

The fuel injection camshaft contains one lobe per cylinder. Its operation is timed in a similar manner to the breaker cam in an old-style ignition distributor, and its position is similarly advanced or retarded by centrifugal weights (discussed under TIMING). In most automotive applications, the camshaft is located in a separate pump housing, along with the plunger pumps as shown in the illustrations in this chapter.

However, some other commercial diesels include the camshaft and plunger pumps within the cylinder head assembly.

Roller Tappets

Roller tappets are used to reduce friction and wear. Each roller tappet rides on its own camshaft lobe, rising and falling with the contour of the cam (see Figure 9–7).

Plunger Pumping Elements

The lower end of the fuel plunger resembles the stem of ordinary intake and exhaust valves. Both have return springs; in the case of the

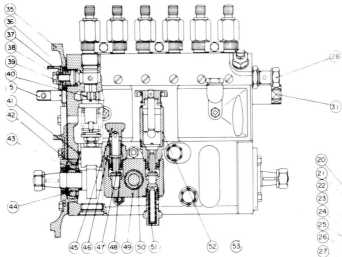

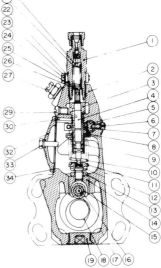

1	Delivery valve holder	15	Roller	29	Screw	43	Tapered roller bearing
2	Pump housing	16	Roller bushing	30	Pin	44	Adjusting shim
3	Plunger barrel	17	Roller pin	31	Cap	45	Distance ring
4	Plunger	18	Camshaft	32	Plate cover	46	Connector bolt
5	Rack guide screw	19	Screw plug	33	Cover screw	47	Nipple
6	Control rack	20	Delivery valve spring	34	Gasket	48	Check valve
7	Control pinion	21	Lock plate	35	Governor housing	49	Check valve spring
8	Upper spring seat	22	Lock washer	36	Screw plug	50	Feed pump housing
9	Control sleeve	23	Bolt	37	Spring eye	51	Piston
10	Plunger spring	24	Delivery valve	38	Lock washer	52	Connector bolt
11	Lower spring seat	25	Delivery valve seat	39	Set screw	53	Priming pump
12	Adjusting bolt	26	Air bleeder screw	40	Control rack	54	Drain plug
13	Lock nut	27	Delivery valve gasket	41	Pinion clamp screw	55	Bearing cover
14	Tappet	28	Gasket	42	Tappet guide		

FIGURE 9-4. **A fuel injection pump (*Courtesy of Nissan Diesel Motors Ltd./Marubeni America Corporation*).**

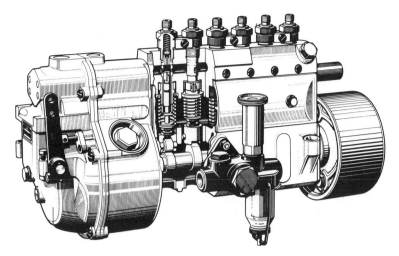

**FIGURE 9-5.   Access to the fuel gallery (*Courtesy of Robert Bosch Corpora-
tion*).**

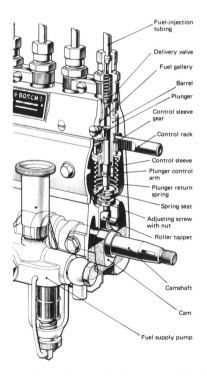

Fuel-injection
tubing

Delivery valve

Fuel gallery

Barrel

Plunger

Control sleeve
gear

Control rack

Control sleeve

Plunger control
arm

Plunger return
spring

Spring seat

Adjusting screw
with nut

Roller tappet

Camshaft

Cam

Fuel supply pump

**FIGURE 9-6.   A cross-section of a multi-plunger fuel injection pump (*Cour-
tesy of Robert Bosch Corporation*).**

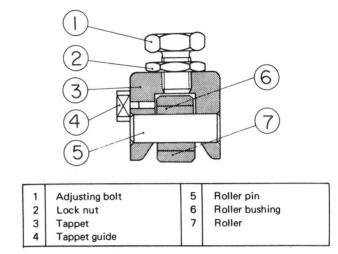

1	Adjusting bolt	5	Roller pin
2	Lock nut	6	Roller bushing
3	Tappet	7	Roller
4	Tappet guide		

FIGURE 9-7. A multi-cylinder injection pump roller tappet (*Courtesy of Nissan Diesel Motors Ltd./Marubeni America Corporation*).

plunger, the spring pushes the plunger and tappet back down after the cam lobe rotates past a high spot.

The upper end of the plunger is where the pumping action takes place. Each pumping element contains the plunger itself and a barrel in which the plunger slides up and down. The parts are machined to very close tolerances; if one fails, both must be replaced as a unit.

Figures 9-8 and 9-9 show various cross sections of a typical plunger and barrel assembly. Notice that:

1. The barrel contains two holes, one on either side. These holes let fuel from the gallery enter the space over the plunger. The holes also let unused fuel return to the galleries.

2. The plunger itself contains a vertical slot cut from the top down one side and a helical cut that joins with the vertical cut.

When the plunger rises (due to the action of the cam lobe and tappet), it squeezes fuel in the top of the barrel toward the delivery valve. The actual stroke or travel of the plunger is the same for all engine speeds. However, the amount of fuel delivered varies according to the rotational position of the plunger.

To see how this works, examine the position of the plunger's vertical and helical cuts as pictured in Figures 9-8 and 9-9. Notice that the plunger is rotated by the control rack.

At maximum fuel delivery, the helical cut does not line up or register with the hole in the barrel unit after the plunger has gone almost all the way up the barrel. At reduced delivery, the helical cut lines up sooner. And at no delivery, the vertical slot faces the hole before the plunger even begins to rise.

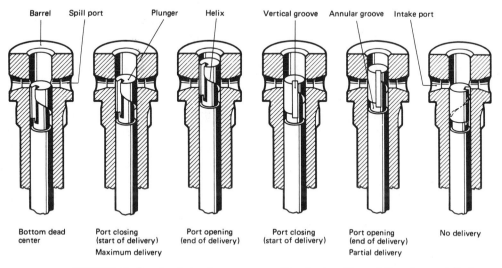

FIGURE 9–8. **An injection pump with dual orifices** (*Courtesy of Robert Bosch Corporation*).

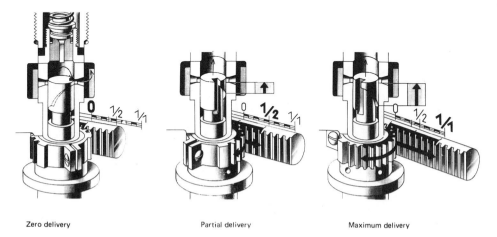

FIGURE 9-9. **A turning mechanism for controlling fuel delivery** (*Courtesy of Robert Bosch Corporation*).

So, at the maximum delivery position, the fuel in the barrel over the plunger is subjected to the greatest pumping pressure for the longest period of time. The fuel isn't allowed to escape back into the gallery until after the plunger has had almost a full upward stroke. At the other positions, because of the relation of the helical or vertical cuts with the barrel holes, the fuel returns earlier to the gallery. (In any of these cases, the fuel pressure over the plunger is greater than the pressure in the gallery. Whenever a path is provided between the two regions, the fuel will always flow back to the gallery.)

Therefore, even though the stroke of the pump remains the same, the EFFECTIVE stroke and the amount of fuel delivered varies. It depends, as we've seen, on the rotational position of the plunger with respect to the barrel openings.

The plungers illustrated in the previous figures are called dual-orifice units. That is because they contain two holes in the barrel. The pumping elements pictured in Figures 9–10 and 9–11 are called single-orifice fuel injectors because they only contain one opening in the barrel. Notice that the vertical slot has been replaced by a hole drilled down through the top of the plunger. This hole leads to an angular slot cut into the side of the plunger. The slot performs the same function as the helical slot in dual-orifice injectors.

Control Rack

The teeth of the control rack engage teeth on control sleeves connected to the plungers. As the rack goes back and forth, it moves all the plungers in unison. The rack's position, or how far it moves, is determined by the accelerator pedal acting in combination with the governor.

Governor

If it weren't for the unique ability of diesel engines to destroy themselves, a governor would not be needed. A simple, direct linkage between the control rack and the accelerator pedal could be used. However, *if* that sort of arrangement were attempted, the self-igniting diesel engine would accelerate explosively at the first low-load or no-load condition. Using just the fuel contained in the combustion chamber at that moment, the engine would run up to 120,000 RPM—assuming it didn't fly apart first, which is more likely. Consequently, the governor is needed to sense engine speed and push against the control rack to reduce fuel flow if the engine attempts to speed up on its own.

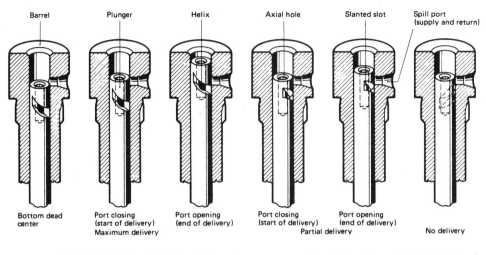

FIGURE 9-10. **Single orifice plunger operation (***Courtesy of Robert Bosch Corporation***).**

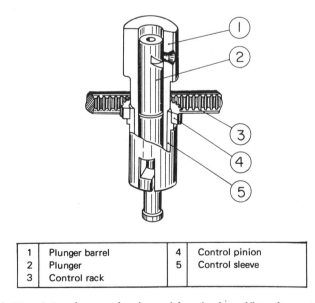

1	Plunger barrel	4	Control pinion
2	Plunger	5	Control sleeve
3	Control rack		

FIGURE 9-11. **A turning mechanism with a single-orifice element (***Courtesy of Nissan Diesel Motors Ltd./Marubeni America Corporation***).**

Various kinds of governors are used in different types of diesel engines. The kind most often found in automotive fuel Bosch injection units is called a minimum/maximum mechanical system. It limits fuel delivery at the upper and lower ends of the speed range. It also, and most importantly, prevents runaway speeds within the operating range. The operator's foot on the accelerator pedal still controls the particular speed. The governor simply makes sure the speed stays where it is supposed to be.

Figure 9–12 is a schematic showing the basic operating principles of a maximum/minimum mechanical governor. The accelerator pedal (I) is connected by a throttle linkage (H) to a throttle sleeve (G). As the pedal moves back and forth, the motion from the throttle sleeve is transmitted via a tension spring (F) to a sliding sleeve (C). The motion of the sliding sleeve is transmitted in turn by another linkage (D) to the rack rod (E). Stops (K and J) at both ends of the rod control the range of the rod's travel and hence, maximum and minimum fuel delivery.

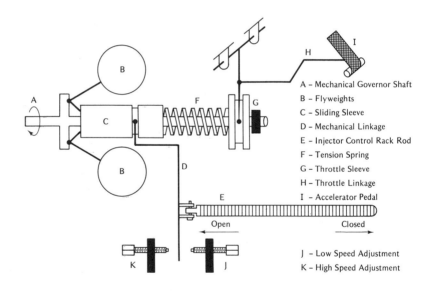

A – Mechanical Governor Shaft
B – Flyweights
C – Sliding Sleeve
D – Mechanical Linkage
E – Injector Control Rack Rod
F – Tension Spring
G – Throttle Sleeve
H – Throttle Linkage
I – Accelerator Pedal

J – Low Speed Adjustment
K – High Speed Adjustment

FIGURE 9–12.   Basic centrifugal governor.

Two types of full throttle stops (for controlling maximum delivery) are shown in Figures 9–13 and 9–14. Both are adjustable; however, the one pictured in Figure 9–14 contains a spring stop so that additional fuel for starting can be obtained by pressing harder on the accelerator. This overcomes the spring tension and extends the travel of the control rack.

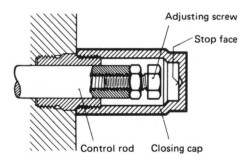

FIGURE 9-13. **Fixed control rod stop (*Courtesy of Robert Bosch Corporation*).**

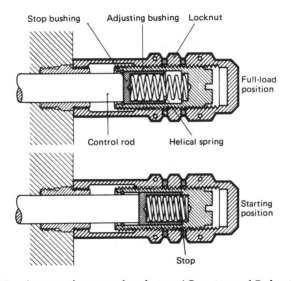

FIGURE 9-14. **Automatic control rod stop (*Courtesy of Robert Bosch Corporation*).**

In a gasoline engine, the components just described would be adequate to determine engine speed. But, as indicated previously, the diesel needs additional controls to prevent the engine from going too fast. This control is provided by the flyweights (B) mounted on the shaft (A) (Figure 9-12).

The shaft is driven either directly by the engine or by the injector pump camshaft. In both cases, the shaft speed is directly proportional to the engine speed. So, as the engine speeds up, the flyweights swing out due to centrifugal force. Then, as the engine slows down, the weights swing back. The action of the flyweights causes the sliding sleeve to move back and forth. That, in turn, changes the position of the control rack and the amount of fuel delivered.

Therefore, two forces are always working on the sliding sleeves and control rack: (1) the pressure of the operator's foot on the accelerator pedal and (2) the action of the flyweights. In effect, the control rack "floats," moving back and forth in response to both forces.

It might seem that the operator of a diesel engine must constantly make compensating throttle adjustments. However, that is not the case. The governor acts almost instantly as changes in load (or other factors) cause the engine to suddenly speed up. While the operator maintains a steady throttle setting for a given engine speed, the flyweights, control sleeves, and control rack move back and forth to control the amount of fuel required to make sure the engine doesn't run away.

ADDITIONAL BOSCH PUMP FEATURES

The last items to be discussed before going on to fuel injection lines and nozzles are: (1) timing, (2) fuel feed pumps, and (3) delivery valves.

Timing

Just as the spark of a gasoline engine must be advanced or retarded as engine speed changes, the moment that fuel is injected must be varied in diesel engines. This is accomplished in many Bosch systems by a set of flywheel weights mounted on the end of the injector pump camshaft. These weights act in a similar manner to the weights in the mechanical advance systems used in pre-solid state, gasoline engine ignition distributors. When the engine speeds up, the weights fly out. That action is transmitted via a linkage system to the injection camshaft, causing it to rotate into an advance position. In other words, the camshaft is rotated so the lobes lift the tappets sooner. The reverse takes place when the engine slows down.

Fuel Feed Pump (and Primer Pump)

The fuel feed pump forces fuel at a pressure of about 14–15 PSI through the filter to the injector gallery. Some feed or supply pumps are electrically operated. Many, like the one shown in Figure 9–15, are mounted directly on the side of the injector pump and are driven by an eccentric on the injector pump camshaft.

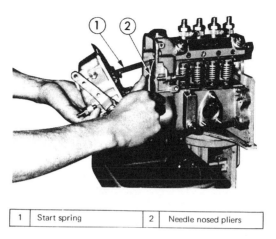

| 1 | Start spring | 2 | Needle nosed pliers |

FIGURE 9-15. A fuel feed pump (*Courtesy of Nissan Diesel Motors Ltd./Marubeni America Corporation*).

The primer pump screwed into the side of the feed pump supplies fuel to the gallery during start-up. It can also be hand-operated for line filling and venting purposes after repairs have been made to the injector system.

Fuel Delivery Valves

Each fuel plunger unit has a fuel delivery valve sitting on top. When the fuel pressure over the plunger becomes great enough to overcome the tension on the delivery valve spring, the valve opens and fuel passes into the delivery valve tubing. The valve closes when the fuel pressure no longer exceeds the spring tension. The delivery valve components are pictured in Figure 9-16.

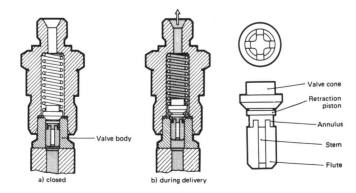

FIGURE 9-16. A delivery valve (*Courtesy of Robert Bosch Corporation*).

DIESEL FUEL INJECTION LINES AND NOZZLES (FOR BOTH TYPES OF SYSTEMS)

The fuel injection lines and nozzles used by both Bosch and distributor pumps are similar and can be discussed together.

Fuel Injection Lines

Fuel injection lines generally fall into two categories, those used for low and those used for high pressure fuel delivery. Low pressure lines conduct fuel from the fuel tank (via the filter) to the feed pump. They are also used to return unused fuel from the filter, pump, and nozzles back to the tank. Low pressure lines are made from steel coated with a zinc or tin alloy. The coating prevents rust buildup that can result from condensation either inside or outside the line.

High pressure lines conduct fuel from the pump to the injector nozzles. Generally they are the same length to minimize differences in timing between one cylinder and another. These steel lines are generally stronger than those in low pressure use. High pressure fittings are used at both ends. The lines and fittings are sensitive to abuse; therefore, special care (as outlined in shop manuals) must be taken when they are serviced.

INJECTION NOZZLES

In a gasoline engine, all the complex activities of the ignition system come together to produce an electrical arc at the tip of the spark plugs. The same thing is true in a diesel engine for the injection nozzles; they serve as a focus for the operation of the entire system. In the simplest sense, an injector nozzle need not be anything more than a tube projecting into the combustion chamber. However, when viewed as a part of the entire injection system, the nozzle requirements become more complex. Following are some factors that must be considered.

Size and Calibration.   Although many injector nozzles might appear alike, there are actually subtle differences in the nozzles used in various types and sizes of engines. Nozzles must be sized and calibrated according to the precise requirements for a particular type of installation. Consequently, random exhange of nozzles between different systems should not be attempted.

Injection Termination. After the desired quantity of fuel has been injected, output from the nozzle must cease instantly. Any dribble from the nozzle tips will result in excessive fuel consumption, poor performance, and excessive smoking. This control is provided both by the nozzle and the delivery valve(s) and will be discussed after this section.

Proper Spray Pattern. For proper burning, fuel must be injected from the nozzle tips in the proper pattern. This is determined by the injection pressure, fuel used, and the design of the nozzle tip.

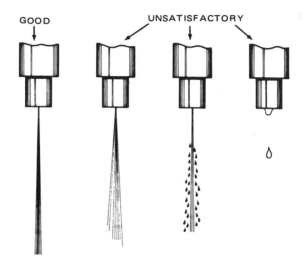

FIGURE 9-17. Nozzle spray patterns (*Courtesy of Nissan Diesel Motors Ltd./Marubeni America Corporation*).

Components

In order to satisfy these considerations, most injector nozzles employ the following major components as pictured in Figure 9-18:

1. body
2. inlet/inlet passage
3. needle valve
4. pressure chamber
5. nozzle tip/spray orifice
6. spring
7. return line

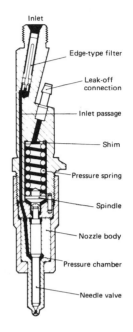

Inlet

Edge-type filter

Leak-off connection

Inlet passage

Shim

Pressure spring

Spindle

Nozzle body

Pressure chamber

Needle valve

FIGURE 9-18. A cross-section of a Bosch fuel injection nozzle (*Courtesy of Robert Bosch Corporation*).

Operation

Fuel under pressure from the injector pump goes through the inlet/passage into the pressure chamber. As the chamber fills up, the fuel pushes on the exposed annular area of the needle valve. When the pressure becomes great enough to overcome the tension on the spring, the needle lifts away from the valve seat, allowing the fuel to spray out of the orifice into the combustion chamber. Then as the pressure diminishes, the spring pushes the needle back down to close the needle seat and stop fuel delivery. Excess fuel over the needle, in the spring area, goes back to the tank through the return line.

Two Types of Nozzles

Two types of nozzle tips are in common use, the "hole" type pictured in Figure 9-19 and the pintle type shown in Figure 9-20. The "hole" type nozzle tip, as the name implies, is a simple hole in the end of the nozzle. The pintle nozzle provides a pintle or rod-like element that projects through a hole in the tip of the nozzle.

The pintle is an extension of the needle valve and is lifted, along with the valve, each time injection takes place. This action provides a degree of self-cleaning by preventing carbon buildup on the end of the injector. Pintle nozzles are found primarily in engines using pre-combustion or turbulence combustion chambers.

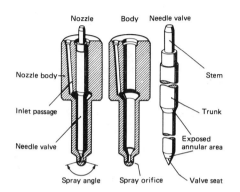

FIGURE 9-19. **A hole-type fuel injection nozzle (***Courtesy of Robert Bosch Corporation***).**

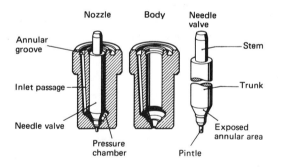

FIGURE 9-20. **A pintle-type injection nozzle (***Courtesy of Robert Bosch Corporation***).**

RELATIONSHIP BETWEEN DELIVERY VALVES AND NOZZLE VALVES

The pump delivery valves and the nozzle needle valves work together to keep the lines full of fuel at all times and to make sure that fuel shut-off is instantaneous.

As noted before, precise shut-off is necessary to prevent dribbles that can result in reduced power and economy and increased smoking.

It is equally important to maintain a solid charge of fuel in the lines. Otherwise air would enter—having the same effect on the diesel fuel hydraulic system as it would have on a hydraulic brake system. Except in this case, instead of mushy brakes, we would have "mushy," uneven fuel delivery.

In a way, the fuel in the lines must act almost like a solid rod. The push from the injector pump on the delivery valve end of the "rod" must have an equal and almost instant reaction at the nozzle valve. Then when the push falls below certain levels, both valves must again act almost instantly to cleanly snip off ends of the fuel rod—before any air enters.

There is also another aspect of the relationship between the two valves to be considered: As fuel moves through the lines, it builds up momentum. When the nozzle valve closes, the force of this momentum tends to keep pushing on the valve. Enough pressure is created to cause leaks and dribbles into the combustion chamber.

To reduce the pressure so that can't happen, a retraction piston (see Figure 9-16) is formed into the lower portion of each delivery valve. In the instant before the delivery valve fully closes, the retraction piston sucks a small amount of fuel out of the line. Not enough fluid is withdrawn to let air enter, only the fraction needed to lower the line pressure below the opening pressure of the nozzle valve.

CLEANLINESS

When servicing or replacing injector nozzles (or when working on any part of the fuel injection system) cleanliness is of vital importance. Dirt or debris will seriously affect engine performance.

DISTRIBUTOR TYPE FUEL INJECTION

As noted before, a distributor fuel injection system is like a gasoline engine ignition distributor. Just as the distributor routes high voltage surges from a single coil to a number of cylinders, a distributor fuel injector routes fuel from a single pumping element to a number of cylinders.

The distributor pump system is used by various companies including GM and Volkswagen. At the time of this writing, GM buys its units from Roosa Master, a longtime manufacturer of distributor pumps for truck and industrial applications.

Basic Operating Groups and Principles

Figure 9–21 is a mechanical schematic showing the basic operating groups of a typical distributor pump. We'll use this simplified illustration to acquire a general understanding of how the system works. Then we'll look at detailed illustrations of the same components for a more thorough description of the operation.

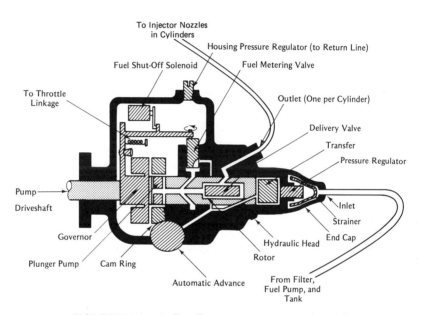

FIGURE 9–21.   A distributor type pump schematic.

First, notice that the driveshaft and rotor run the length of the assembly. These components are driven by the engine camshaft and turn at one-half crankshaft speed. Rotating with the driveshaft/rotor assembly are the: (1) governor weights, (2) plunger pump, (3) delivery valve, and (4) feed pump. The hydraulic head, located at the right end of the illustration, provides a bore for the rotor as well as a number of fuel passages to different locations. Fuel flows through the system in the following manner.

1. A fuel pump pulls the fuel from the tank through the fuel filter.
2. A vane type transfer pump at the end of the rotor pulls the fuel from the filter, past the injector inlet and through a strainer.

3.   From the transfer pump, the fuel goes through a passage in the rotor to a channel in the head. The force of the fuel leaving the feed pump is sensed by a pressure regulator. It routes part of the fuel back to the inlet side of the pump (thereby controlling the outlet pressure).

4.   Once in the hydraulic head, the fuel is directed to two channels. One dead-ends at the hydraulically operated advance mechanism.

5.   The other goes behind the rotor to a fuel metering valve.

6.   The amount of fuel flowing through the metering valve is controlled by the rotational position of the cut-away portion on the lower part of the valve. The position changes as the valve is rotated back and forth in response to pressure from the governor and the accelerator pedal.

7.   Fuel goes from the metering valve into a series of charging channels surrounding the rotor.

8.   Two of these channels line up with two inlet passages in the rotor.

9.   The two inlet passages lead to a plunger pump. It consists of two plug-like plungers that slide back and forth in holes in the rotor. Fuel under pressure from the transfer pump forces the two plungers apart.

10. Surrounding the circular path traveled by the plungers is an internal cam ring (cam surfaces located on the inside of the ring). As the plunger rollers strike the cam surfaces (one per cylinder), the plungers are pushed together. This forces the fuel under pressure back into a rotor channel leading to the delivery valve.

11. Fuel goes from the delivery valve through a channel in the rotor to a passage in the hydraulic head. This passage directs fuel to the outlet connected to a nozzle line. The head passage adjoins the delivery valve channel once per rotor revolution. Similar head passages direct fuel to the other cylinders in the engine.

12. As the rotor turns, the cycle repeats for every cylinder in the engine. The various channels that honeycomb the rotor line up with corresponding passages in the hydraulic head, move apart, then line up again on the next revolution of the rotor.

> **NOTE**
>
> **1.** Fuel delivery is advanced or retarded by the automatic advance. It shifts the position of the cam ring thereby causing the plungers to come together earlier or later. The advance mechanism operates by hydraulic pressure from the feed pump—responding to pressure changes as the engine speed (and hence the pump pressure) changes.
>
> **2.** Runaway engine speed is prevented by the governor mechanism. It contains centrifugally operated weights that move a linkage to adjust the position of the metering valve.

THE COMPONENTS IN MORE DETAIL

The next paragraphs provide additional facts about some of the previously introduced components as well as information about some components not shown before. Refer first to Figure 9–22, then to the other figures as directed.

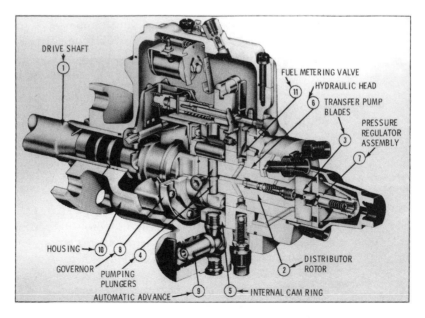

FIGURE 9-22. A fuel injection pump (*Courtesy of Oldsmobile Division, General Motors Corporation*).

Driveshaft

Several means may be used to turn the injector pump driveshaft. GM uses a gear assembly running off the rear of the camshaft sprocket as shown in Figure 9–23. Volkswagen uses the same timing belt that drives the camshaft to rotate the injector pump driveshaft.

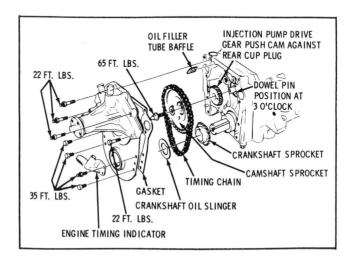

FIGURE 9–23.   A General Motors Corporation injection pump "drive gear" (*Courtesy of Oldsmobile Division, General Motors Corporation*).

Rotor

The driveshaft and rotor are separate parts, connected by a slot and tang assembly.

Pressure Regulator

The pressure regulator shown in Figure 9–24 controls the force of fuel leaving the transfer pump. It does this by routing some of the fuel back to the inlet side of the pump. The main components are a piston, spring, and regulating ports. As fuel leaves the transfer pump, it pushes against the piston and spring. When the spring is compressed sufficiently, the edge of the piston uncovers the regulator port, providing a path for fuel to return into the inlet side of the transfer pump. To prevent excessive transfer pump pressure at higher speeds, a relief slot is provided near the extreme end of piston travel.

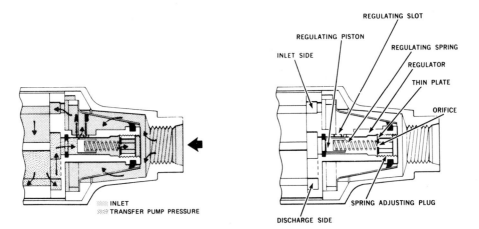

FIGURE 9-24. The operation of the regulator assembly (*Courtesy of Oldsmobile Division, General Motors Corporation*).

Transfer Pump

The transfer pump is machined into the right end of the rotor shaft as shown in Figure 9-25. Vanes (not shown) fit into the four slots at the end of the shaft. A circular ring (also not shown yet) surrounds the vanes. The inside of the ring is off center (eccentric) with respect to the shaft so that the vanes move in and out as the shaft turns.

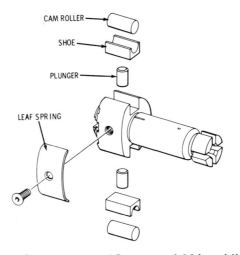

FIGURE 9-25. Transfer pump parts (*Courtesy of Oldsmobile Division, General Motors Corporation*).

Figure 9-26 diagrams the operation of the pump. As the rotor turns, the volume between the vanes increases and decreases, thus increasing and decreasing the pressure in space between the vanes. The low pressure region adjacent to the inlet port pulls fuel through the strainer into the pump. Then, as the rotor goes on around, the space containing the fuel becomes a high pressure region. At the point of maximum pressure, fuel is forced through the outlet port.

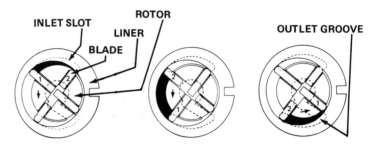

FIGURE 9-26.   Transfer pump operation (*Courtesy of Oldsmobile Division, General Motors Corporation*).

Plunger Pump/Cam

The left side of the rotor pictured in Figure 9-25 shows the parts of the plunger pump. An end view of these parts, plus the cam ring is shown in Figure 9-32. As noted before, fuel goes from the metering valve into the space between the plungers. The amount of fuel admitted for any charging cycle determines how far apart the plungers move. At idle or lower speeds, the plungers don't move apart as far as they do at higher speeds. Maximum plunger travel and maximum fuel delivery is limited by a single leaf spring that contacts the ends of the roller shoes. Figure 9-27 shows the plunger pump being charged. Figure 9-28 shows the pump forcing fuel through the delivery valve.

Delivery Valve

The single delivery valve of the distributor injector is similar to the multiple valves used in the Bosch system. The principal components of the valve are a piston, spring, and outlet port. When the fuel pressure from the plungers becomes great enough, the piston uncovers the outlet passage. Then when the pressure drops off, the piston closes the opening. As in the Bosch system, the delivery valve and its nozzle valves work together to make sure that fuel shut-off is instantaneous and that the high pressure lines are filled with fuel at all times.

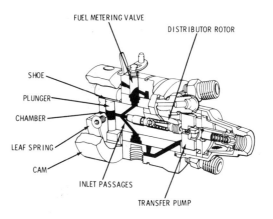

FIGURE 9-27. The charging cycle (*Courtesy of Oldsmobile Division, General Motors Corporation*).

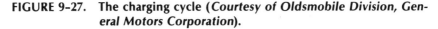
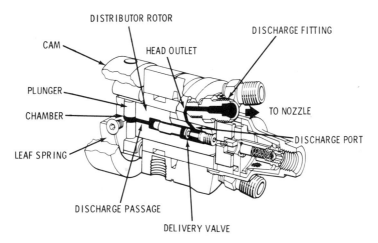

FIGURE 9-28. The discharge cycle (*Courtesy of Oldsmobile Division, General Motors Corporation*).

Fuel Return Circuits

Before arriving at the metering valve, fuel goes through a vent passage in the hydraulic head, as pictured in Figure 9-29. Located in the passage is a vent valve consisting of a small wire in a calibrated opening. Air and excess fuel from the transfer pump pass through the vent into the governor linkage compartment. From there, the bubbly air and fuel mixture flow through a pressure regulator in the governor compartment, then through a return line to the fuel tank.

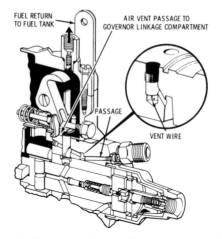

FUEL RETURN
TO FUEL TANK

AIR VENT PASSAGE TO
GOVERNOR LINKAGE COMPARTMENT

PASSAGE

VENT WIRE

FIGURE 9-29.   The fuel return circuit (*Courtesy of Oldsmobile Division, General Motors Corporation*).

Governor

The governor in a distributor type pump serves the same purpose as the governor in a Bosch system—it maintains engine speed in the face of varying load conditions. As before, the governor consists of centrifugally operated weights that transmit engine speed changes into force acting on a fuel control mechanism. And, as before, the force of the governor is balanced by the force of the driver's foot working through a linkage system to the fuel control.

In the example shown in Figures 9-30 and 9-31, the weights ride loosely on the governor thrust sleeve. As the sleeve rotates, the heavy ends of the weights tip outward due to centrifugal force. However, the other ends of the weights are restricted by the weight retainer. Consequently, the tipping action of the weights causes the thrust sleeve to slide back.

As the sleeve slides back, it pushes on the governor arm. The arm in turn pivots on the knife edge of the pivot shaft. This action of the governor arm is transmitted to the linkage shaft, which rotates the metering valve.

The force of the weights acting on the sleeve and arm is balanced by the governor spring. And the pre-load on the spring is manually controlled by the throttle linkage to which the spring mechanism is attached.

So, just as the control rack in the Bosch system "floats" in response to forces from two sources, the metering valve in the distributor system "floats" in response to forces from the governor mechanism and the accelerator pedal.

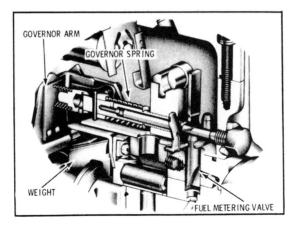

FIGURE 9-30.   A mechanical governor (*Courtesy of Oldsmobile Division, General Motors Corporation*)

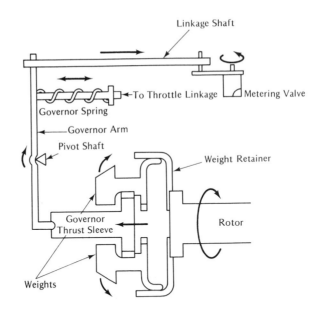

FIGURE 9-31.   A schematic of governor operation.

Fuel Shut-Off Control

Notice in Figure 9-22 the electromechanical device located in the upper part of the governor chamber. This is a fuel shut-off valve. When the ignition switch is turned off, this device pushes the metering valve linkage arm so that no fuel can be delivered through the system.

Automatic Advance

The automatic advance system pictured in Figure 9–32 shifts the position of the internal cam ring to vary the time that fuel delivery starts. Injecting fuel earlier at higher speeds ensures that combustion takes place at the most effective piston position. This, in turn, ensures the best power output from the least amount of fuel with the least amount of smoke coming from the tailpipe.

The basic components of the advance system include a piston, pin assembly, and spring. The piston is located in a bore in the pump housing. Fuel under pressure from the transfer pump goes into the bore and causes the piston to move against the spring. The motion of the piston is transmitted to the pin, which shifts the cam ring opposite the direction of rotor rotation. A reed check valve traps fuel in the piston chamber, thereby restricting the natural tendency of the cam ring to return to the retard position at the moment of injection. The fuel in the piston chamber is allowed to bleed through a hole (actually a control orifice) in the piston.

The amount of advance is proportional to the transfer pump pressure, which is proportional to the engine speed. Maximum advance is limited by the length of the piston. The moment advance starts is controlled by the pre-load force on the advance spring, which is controlled by a trimmer screw (adjusted at the factory).

Injection is retarded when the engine slows down and the pump pressure diminishes. Then, the spring pushes the advance piston, which, working through the pin assembly, moves the cam ring back around.

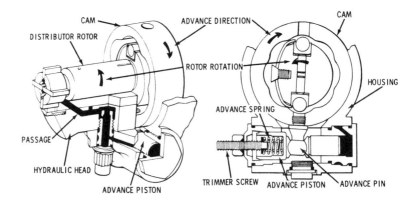

FIGURE 9–32. An automatic advance system (*Courtesy of General Motors Corporation*).

CHAPTER 10

Diesel Cooling Systems

INTRODUCTION

Diesel engine cooling systems are virtually identical to those used in gasoline engines, particularly those in high performance or heavy duty gasoline powered vehicles. This chapter briefly reviews the operating principles of cooling systems and notes some of the major components.

DEALING WITH HEAT

Heat is a critical factor in the operation of any internal combustion engine:

1. The burning process yields waste heat, which must be dissipated.
2. The engine operates best within a certain heat/temperature range, neither too cold nor too hot.

3. Although not direct parts of the internal combustion engine, the driver and passengers require heat for satisfactory winter performance.

The cooling system deals with these considerations. It gets rid of excess heat, maintains engine temperature in a certain range, and routes engine heat into the passenger compartment for wintertime comfort.

FUNDAMENTALS OF OPERATION

The cooling system operates by circulating a coolant liquid under pressure through the engine and radiator. Heat is conducted to the liquid in the engine, then radiated into the atmosphere by the radiator. The temperature of the liquid (and hence, the engine) is maintained within a certain range by a thermostat valve. When the engine gets too hot, the valve directs more coolant to the radiator where the heat is radiated away. When the engine is too cold (e.g., at start-up) the valve causes most of the liquid to circulate through the engine where very little heat escapes.

COOLANT FLOW AND MAIN COMPONENTS

The following paragraphs summarize the flow of coolant through the main components in a typical cooling system.

1. *From Water Pump.* Although the coolant circulates constantly with no real starting point, for the sake of discussion, we can say the flow begins at the water pump. Most water pumps are of the impeller type, located at the front of the engine and driven by a belt running off the crankshaft. The center part of the propeller-like impeller is a low pressure region drawing water from the pump inlet. Once the water enters the pump, centrifugal force throws it toward the tips of the blades and the outlet (see Figure 10–1).

2. *Into the Water Jackets.* From the pump, the coolant flows into the water jackets or cooling channels cast into the block and head. The jackets are designed to bring the liquid into contact with the hottest parts of the engine (see Figure 10–2). The coolant begins its passage through the engine at the lower part of the block. As the liquid moves upward in the engine, it becomes hotter. Heated liquids tend to rise (like balloons), which makes the job of the pump easier.

3. *Into the Heater Coils.* Some of the heated liquid passes from
the water jacket, through rubber hoses into the heater coils.
Air blowing over these coils picks up heat and carries it to the
passenger compartment. The liquid is returned from the coils
back to the cooling system.

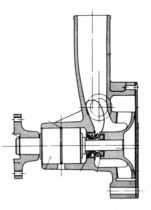

FIGURE 10-1. **A cross-section of a water pump (*Courtesy of Nissan Diesel
Motors Ltd./Marubeni America Corporation*).**

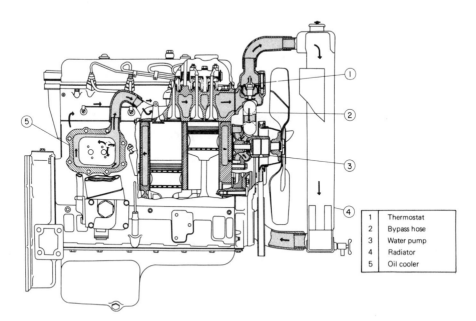

1	Thermostat
2	Bypass hose
3	Water pump
4	Radiator
5	Oil cooler

FIGURE 10-2. **Cooling system passages in a Nissan 4-cylinder engine
(*Courtesy of Nissan Diesel Motors Ltd./Marubeni America
Corporation*).**

4. *To the Thermostat Valve.* After coolant has circulated through the entire engine, it comes to the thermostat valve (usually located at the top front of the head or block). Most modern engines use a pellet-type thermostat. The pellet is a combination wax and copper sheath. It surrounds an operating rod connected to a valve. When the pellet heats up, it expands and pushes the rod and valve into an open position. When the pellet cools, the opposite action takes place to close the valve.

As noted before, the purpose of the thermostat is to direct coolant to the radiator when the engine is hot and keep it out of the radiator when the engine is cool. The question then arises, how does the coolant get back to the engine when the thermostat blocks the flow to the radiator (see Figure 10–3).

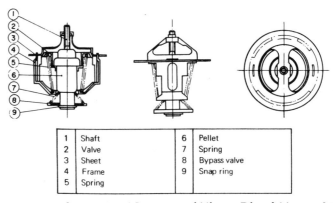

1	Shaft	6	Pellet
2	Valve	7	Spring
3	Sheet	8	Bypass valve
4	Frame	9	Snap ring
5	Spring		

FIGURE 10-3. A thermostat (*Courtesy of Nissan Diesel Motors Ltd./Marubeni America Corporation*)

5. *To the Bypass Valve.* The answer in many cases is a bypass valve that allows coolant to flow back to the water pump, where it is directed through the engine again. Several types of bypass valves are used. One type is operated by a spring weaker than the thermostat. Whenever the latter closes the former opens. Another type of bypass valve is always open, allowing a certain amount of liquid to flow through the engine at all times.

6. *Through the Top Radiator Hose.* When the engine is hot, most, if not all of the liquid is directed from the thermostat into a hose connecting the top of the engine to the top of the radiator. This hose is often made of molded rubber and it may be (but isn't necessarily) reinforced with a wire insert.

7. ***Into the Radiator.*** Once inside the radiator, the hot coolant is forced from the top to bottom through small tubes or passageways. These passageways are connected to a honeycomb-like grid of thin metal baffles. Heat is conducted from the passageways to the baffles, which radiate the heat into the air blowing through the intervening spaces (see Figure 10-4).

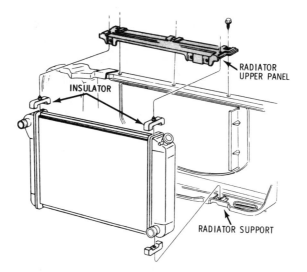

FIGURE 10-4. **Typical radiator assembly (*Courtesy of Oldsmobile Division, General Motors Corporation*).**

The network of cooling channels and baffles are collectively called the cooling core. The reservoir on top where the liquid enters is called the top tank. The bottom reservoir is, naturally, the bottom tank.

The top tank contains the pressure cap and an overflow line to let fluid run off as pressure and temperature go up.

The bottom tank contains a drain valve and connections to the outlet hose (see Figure 10-5).

8. ***Through the Bottom Hose.*** After going down through the radiator, the cooled fluid passes into the bottom hose, usually made out of molded rubber and likely to be reinforced by a wire insert. The insert is necessary because the coolant exerts less pressure when it is cool—so little, in fact, that air pressure sometimes tends to collapse the bottom hose.

9. *Back to the Water Pump.* The bottom hose is often connected to the water pump inlet. At this point, the circuit is complete.

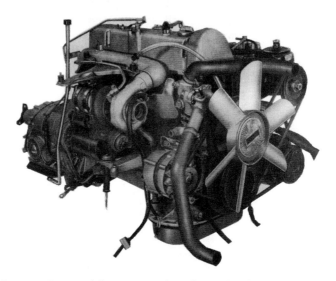

FIGURE 10-5. Top and bottom coolant hoses leading from the engine to the radiator (*Courtesy of Mercedes-Benz*)

AIR FLOW AND COMPONENTS

At normal cruising speeds, enough air flows through the radiator to take the heat away and reduce the temperature of the coolant. However, at slower speeds and when the vehicle is at a standstill, the air flow is either non-existent or insufficient to do the job.

The primary function of the fan is to pull air through the radiator during these times. The fan, which may have two, three, four, or more blades is driven by a belt and pulley running from the crankshaft. In many cases, the same pulley that operates the fan also turns the water pump.

Since the fan is not required at higher road speeds, a number of modern cars have devices to disengage the fan when it isn't needed. One way of doing this is to drive the fan through a temperature sensitive, fluid clutch assembly. A valve operated by a bi-metal spring controls the flow of drive fluid to the clutch. When the air temperature goes up, the valve directs more fluid to the drive allowing the fan to turn faster and draw more air. The converse happens when the air temperature goes down (see Figure 10-6).

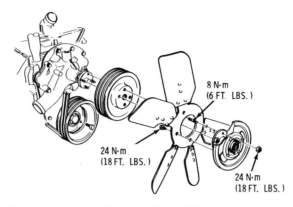

FIGURE 10-6. An example of a fan clutch (*Courtesy of Oldsmobile Division, General Motors Corporation*).

In order to make sure the cooling air goes where it will do the most good, many modern cars are also equipped with cooling shrouds and baffles. These components surround the fan area, increasing the efficiency of the fan and making sure that all the air entering the radiator area flows through the radiator cooling core.

ADDITIONAL COOLING SYSTEM FACTS

Now that the fundamentals of operation, and coolant and air flow have been reviewed, some additional facts need to be briefly covered.

The Coolant Itself

Virtually all modern cars use a mixture of water and ethylene glycol as the coolant liquid. Water by itself would be unsatisfactory because it freezes at too high a temperature. Water also expands with a great deal of force as it goes from the liquid to the solid state. Combine these two facts with even moderately cold climates and you would have a cooling system with no liquids—plus burst pipes, cracked blocks, etc.

Nor can ethylene glycol be used by itself. Although ethylene glycol remains a liquid over a much broader range (−100 °F to +330 °F versus +32 °F to +212 °F for water), it becomes a slush at higher temperatures and does not flow well.

New cars come from the factory with cooling systems already containing water and ethylene glycol, plus rust inhibitors. The exact glycol content that should be maintained depends on the lowest temperature to which the vehicle will be exposed. The lower the temperature the more glycol should be present.

The conventional hydrometer test device is used to determine the protection provided by the coolant. Whenever water is added, the coolant should be checked.

Radiator Pressure Caps

Not only does ethylene glycol lower the freezing point of the coolant, it also raises the boiling point. That is desirable because the heat transfer to the atmosphere is faster when the coolant temperature increases.

The boiling point of a liquid can also be increased by raising the air pressure over the liquid. That is the job of the pressure cap. Besides providing an opening to the system, it also seals off the system, allowing the pressure to build up so that the boiling point of the liquid increases.

However, the pressure cannot build up indefinitely; parts of the system might rupture due to excessive pressure. Therefore a relief valve is usually built into most pressure caps. When the pressure becomes too great, the valve opens, allowing high pressure vapors to escape out the overflow pipe (see Figure 10–7).

FIGURE 10–7.   A pressure radiator cap (*Courtesy of Oldsmobile Division, General Motors Corporation*).

Most pressure caps also contain a second valve to open the system to the atmosphere when the engine cools off. Otherwise a vacuum would form inside the cooling system, possibly resulting in the collapse of components from the force of normal atmospheric pressure.

Closed Cooling Systems

In old style cooling systems, the overflow pipe simply went down the side of the radiator. Any liquid that entered the pipe ran on the ground and was lost. So, even though the cooling system might have a sealing pressure cap, it wasn't really "sealed." Coolant lost had to be replaced.

Most modern systems use a tank or reservoir to capture the coolant. As the coolant heats up and expands, it flows through the overflow pipe into the tank. The expansion is contained by a cushion of air over the tank. Although fluid can still be lost from the expansion tank, the amount and possibility are minimized.

CHAPTER 11

Diesel Lubricating System

INTRODUCTION

Diesel engine lubricating systems are essentially the same as those used in gasoline engines. The principal differences, as usual, relate to the stresses imposed by diesel operation. The lubricating oils must conform to special diesel requirements. And the lubrication system components must be able to supply oil in adequate amounts and at adequate pressures to parts operating under adverse conditions.

This chapter reviews some basic lubrication concepts and components and examines the unique aspects of diesel lubrication.

FRICTION

Friction can be the enemy or ally of mechanical systems; it depends on their function. Brakes, for instance, couldn't operate unless there was resistance to sliding contact between the brake shoes (or pads) and the drums (or rotors). Friction absorbs the energy of the car's motion, changing it into low grade heat, dissipated by the conductive surfaces of the brake assembly.

However, friction is the decided enemy of most engine parts that slide against one another. Even the smoothest surfaces contain microscopic irregularities that tend to catch and cause resistance to motion.

The way to reduce destructive friction is to prevent parts from touching: that is the primary job of lubricating oils.

ACTION OF LUBRICATING OILS

Lubricating oils form a wedge or barrier between two moving surfaces. In a sense, the oil becomes a two-layer "sandwich." One oil layer or half of the sandwich adheres to one moving surface and the other oil layer adheres to the other surface. Sliding contact is restricted to the motion between the two layers of oil—with the adjacent molecules of oil acting like tiny ball bearings. Although friction still exists due to the cohesion between the oil molecules, it is much less than the friction that would result from dry surfaces rubbing together.

The lubricating action between a shaft and a bore is similar, except that the wedging action is emphasized. Oil that is constantly introduced into the space between a bore and rotating shaft (as in a crankshaft and journal assembly) is shaped into a wedge by the weight and motion of the shaft. This oil wedge lifts the shaft into a slightly eccentric position with respect to the bore and separates the shaft from the bore.

OTHER FUNCTIONS OF LUBRICATING OILS

Besides reducing friction, lubricating oils also perform other functions. They keep the engine clean by washing away debris resulting from wear. Oils also help cool the engine by absorbing heat from the combustion process. This later process is so important that some diesels circulate the oil through a portion of the radiator.

TYPES OF OILS

Various designations have been used to identify different types of oils.

SAE Classifications

The Society of Automotive Engineers provides the most common designations for identifying different kinds of oils: SAE 10, 10W, 20, 20W, 30,

40, 10W–30, and 10W–40, etc. These numbers indicate the relative viscosities of oil, SAE 10 being the least viscous and SAE 40 being the most viscous. The multi-grade oils (e.g., 10W–30) retain the properties of the lower numbered oil at lower temperatures and the properties of the higher number oil at increased temperatures.

Although the SAE system is generally used, there are some common misconceptions about what the terms mean. For instance, the W suffix after lower grade oils does not mean weight. It simply means that the oil has been designed especially for winter use. Nor do the SAE numbers correspond to proportional differences in viscosity; SAE 40 oil is not necessarily twice as thick as SAE 20 oil.

ASTM Classifications

The American Society for Testing Materials and the American Petroleum Institute have devised the following designations for rating the service capabilities of lubricating oils:

SA. . .a light duty oil not especially suited for late model vehicles.

SB. . .a heavier duty oil but still not providing the protection required by engines produced after 1963.

SC. . .a still heavier duty oil of the kind commonly used in vehicles of the mid to late 1960s. Provides detergent (washing) capabilities and protects against rust and corrosion.

SD. . .the kind of oil required to maintain the warranties of many vehicles from 1968 until the mid-seventies. Provides better protection than prior oils, especially at higher engine temperatures.

SE. . .at the time of this writing these oils provide the most protection for parts operating in the hostile environment of modern engines.

The preceding ratings are used primarily to describe oils used in gasoline engines. The following designations relate to diesel engines.

CA. . .a light duty oil not used in diesel automobiles or trucks.

CB. . .another light duty oil, but offering increased protection for engines using high sulphur content diesel fuel.

CC. . .a relatively heavy duty oil used in various kinds of diesel engines.

CD. . .a heavy duty oil used in high speed diesel engines, e.g., in cars and trucks.

WHICH KIND OF OIL TO USE IN A DIESEL

Most diesel manufacturers recommend oils specifically approved for diesel operation. GM states that only oils containing BOTH an SE and CD rating are to be used in their diesels.

The grade or SAE viscosity rating of the oil depends on the temperature. As a general rule, multi-grade oils are to be avoided except in extreme cold weather operation. They offer less protection in high speed applications than single grade oils.

LUBRICATING SYSTEM FUNCTIONAL GROUPS

Following is a discussion of the main components found in most diesel lubricating systems. These components are similar in function, if not actual design, to their gasoline engine counterparts.

Oil Supply

Oil is usually stored in the lower part of the crankcase. The oil pan serves as the bottom of the reservoir as well as a heat transfer device. The latter function is necessary because oil circulating through the engine absorbs heat. This heat is conducted from the oil to the oil pan and from there to the air moving beneath the car (see Figure 11–1a and b).

The crankcase storage method is called wet sump lubrication. Only the oil that has actually entered the lubrication lines and passages is under pressure. In the dry sump system (usually only found in more expensive and/or high performance vehicles), all the oil is under pressure all the time. The dry sump system is comparable to the coolant system where the liquid is stored under constant pressure in the coolant passages within the engine and inside the radiator.

Oil Pick-Up

In the wet sump system, oil is drawn from the crankcase into a screened pickup located beneath the oil surface. In dry sump systems, the oil is always "picked up."

Oil Pump

The oil pick-up is actually an inlet to the oil pump. Several types of pumps are used.

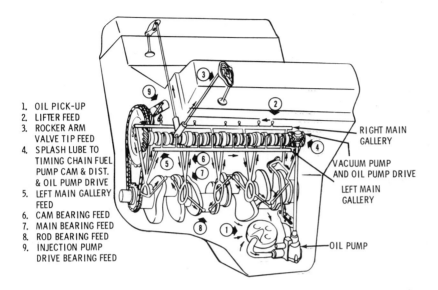

1. OIL PICK-UP
2. LIFTER FEED
3. ROCKER ARM
   VALVE TIP FEED
4. SPLASH LUBE TO
   TIMING CHAIN FUEL
   PUMP CAM & DIST.
   & OIL PUMP DRIVE
5. LEFT MAIN GALLERY
   FEED
6. CAM BEARING FEED
7. MAIN BEARING FEED
8. ROD BEARING FEED
9. INJECTION PUMP
   DRIVE BEARING FEED

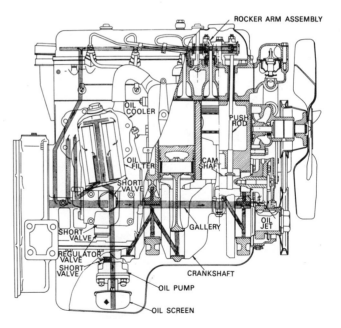

FIGURE 11-1. Lubrication oil passages and oil circuits (*Courtesy of Olds-mobile Division, General Motors Corporation*).

Gear Pump.   This consists of a set of gears located in a housing. As the teeth mesh and unmesh, pressure differences are created drawing the oil past the inlet and pushing it through the outlet.

Rotary Pump.   This consists of two lobed rotors, one turning inside the other. As the rotors turn, the spaces between the lobes increase and decrease. These volume changes create pressure differences: oil is pushed by atmospheric pressure into low pressure regions of the pump. Then, when the low pressure region becomes a high pressure region, the oil is forced out.

Oil pumps are usually turned by a gear and shaft assembly driven by the camshaft (in cam-in-block engines) or by the crankshaft (when the cam is in the head). Wet sump systems usually locate the pump beneath the oil level in the crankshaft. This eliminates any need to prime the pump (see Figures 11–2, 11–3 and 11–4).

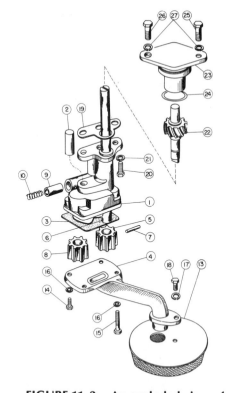

1	Oil pump body	15	Bolt
2	Idler shaft	16	Washer
3	Packing (SD22)	17	Lock washer
4	Oil pump cover	18	Bolt
5	Drive gear	19	Gasket
6	Drive shaft	20	Bolt
7	Pin	21	Lock washer
8	Driven gear	22	Driving spindle
9	Relief valve	23	Driving spindle support
10	Relief valve spring	24	O-ring
11	Washer	25	Bolt
12	Cotter pin	26	Bolt
13	Oil screen	27	Lock washer
14	Bolt		

FIGURE 11-2.   An exploded view of an oil pump (*Courtesy of Nissan Diesel Motors Ltd./Marubeni America Corporation*).

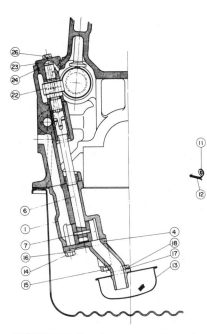

1	Oil pump body	15	Bolt
2	Idler shaft	16	Washer
3	Packing (SD22)	17	Lock washer
4	Oil pump cover	18	Bolt
5	Drive gear	19	Gasket
6	Drive shaft	20	Bolt
7	Pin	21	Lock washer
8	Driven gear	22	Driving spindle
9	Relief valve	23	Driving spindle support
10	Relief valve spring	24	O-ring
11	Washer	25	Bolt
12	Cotter pin	26	Bolt
13	Oil screen	27	Lock washer
14	Bolt		

FIGURE 11-3. A cross-section of an oil pump in the engine block (*Courtesy of Nissan Diesel Motors Ltd./Marubeni America Corporation*).

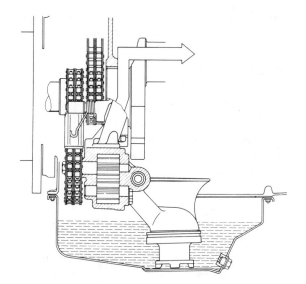

FIGURE 11-4. An oil pump showing the drive chain from the crankshaft (*Courtesy of Mercedes-Benz*).

Oil Filter

From the pump, oil goes to the oil filter. Most modern automotive filters are the full-flow type. All the oil passes through the fibrous filtering
element leaving behind most small particles that were suspended in
the oil. Heavier particles have already dropped to the bottom of the oil
pan.

The filters used in diesel engines are usually similar to those found
in gasoline engines. However, because of the harsher operating conditions of diesels and the stricter cleanliness requirements, manufacturers usually recommend that the filters be changed more often. The
same is true for oil changes (see Figure 11–5).

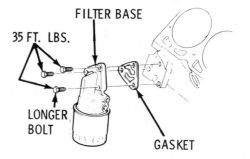

FIGURE 11–5. A fuel filter (*Courtesy of Oldsmobile Division, General
Motors Corporation*).

Oil Coolers

In some diesels, oil goes from the filter to an oil cooler located in the
engine radiator. The cooler transmits engine heat absorbed by the oil
into the atmosphere. In other applications, the cooler may be located
on the engine block and connected to the radiator with hoses similar to
heater hoses (see Figure 11–6).

Oil Passages

From the cooler, if the engine is so equipped, the oil is forced under
pressure to an oil gallery (or galleries in V–8s) running the length of the
block. Each gallery directs oil to the crankshaft, camshaft, valve train,
timing gear, or chain—to virtually any component that needs lubrication. Oil drips off lubricated parts back into the oil pan.

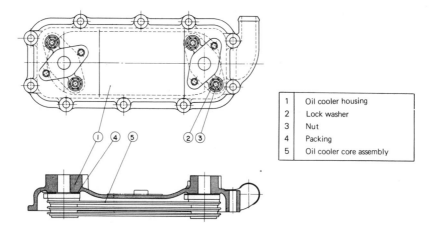

1	Oil cooler housing
2	Lock washer
3	Nut
4	Packing
5	Oil cooler core assembly

FIGURE 11–6.   An oil cooler (*Courtesy of Nissan Diesel Motors Ltd./Maru-beni America Corporation*).

REGULATORS AND BYPASS VALVES

Most diesel lubricating systems are equipped with these two types of valves, both of which control oil pressure, but for different reasons. A regulating valve is usually located between the oil pump and the filter. If the oil pressure exceeds a certain limit (as the engine speeds up), the valve opens allowing some oil to flow back into the oil pan. That way, the pressure is controlled.

A bypass valve senses the pressure that builds up when a critical oil passage is clogged. The valve maintains oil flow to the rest of the engine by routing the oil around the clog. Almost all engines use a bypass valve in the oil filter. This valve opens when the filter is clogged or when the temperature is so low that the oil won't flow properly through the filter. Diesels with oil coolers may also provide a bypass valve to provide an alternate flow path should restrictions occur in the oil cooling system.

DIESEL INJECTOR PUMP LUBRICATION

As noted in Chapter 9, Bosch type pumps are usually lubricated by engine oil taken from the main lubrication system. In a distributor type pump, everything but the pump gear drive is lubricated by the diesel fuel itself. The gear drive is usually lubricated by oil from the engine lubrication system.

OIL PRESSURE GAUGES

Most gasoline powered cars only provide a warning light to indicate low oil pressure. Some diesels also follow this practice. However some offer an oil pressure gauge either as a standard feature or option. In either case, the sensing unit is usually located in an oil gallery where the reading is likely to be most useful.

Oil Pressure

The oil pressure in most diesel engines is between 30 and 50 PSI.

CHAPTER 12

Principles of Electricity and Magnetism

GENERAL

This chapter sets the stage for diesel electrical systems by introducing the separate but related concepts of electricity and magnetism. The first part of the chapter discusses atomic theory, current flow, various kinds of conductors, some practical effects of electrical flow, and the common ways to measure electrical current. The second part of the chapter covers magnetic fields, various kinds of magnets, and a simplified theory of how magnets work. The chapter concludes with brief explanations of solenoids, relays, electric motors, and electric generators.

WHAT IS ELECTRICITY?

First, electricity. We often use the word as if it described a real, physical object. However, that is not exactly what electricity means. How many times have you seen, tasted, or touched electricity? Never, although you may have seen and touched the effects of electricity.

When we talk about electricity, we are actually talking about movement, a flow of tiny particles deep within the heart of matter. The visible or detectable signs of electricity—sparks, shocks, readings on test instruments, etc.—are a result of this flow. When the particles are in motion, they have an effect on the material through which they flow as well as on the surrounding space.

NATURE IN BALANCE

Why do the particles move in a wire, a diesel engine glow plug, an alternator or starter motor, or anywhere else? It has to do with the universal tendency of conflicting forces to try to stay in balance. The planets remain in their orbits because the centrifugal force of their movement is balanced by the force of gravity from the sun. Water stays in a man-made lake because the weight of the water is balanced by the strength of the dam.

All these are examples of forces in balance. However, what happens if something changes, if for instance, the dam breaks? Then the forces are no longer in balance. In the case of the broken dam, water rushes downhill, affecting everything in its path, until the lake is emptied and a new balance is achieved.

It is this urge for balance that causes electrical flow. A battery, alternator, or some other device builds up the number of tiny electrical partricles in one part of a circuit and reduces the number in another part of the circuit. If the circuit is complete, or in other words, if there is an uninterrupted path between the out-of-balance parts, current flows. Excess particles flow to the part of the circuit containing fewer particles. The current continues to flow as long as the battery or alternator maintains the imbalance. And like the water flowing downhill from the broken dam, everything that lies in the path of the current is affected.

ATOMIC STRUCTURE

The next question is, what are these particles? If you have been exposed to the subject previously, you've probably already guessed that the particles are electrons. Electrons, along with protons, neutrons, and other particles, make up atoms, which are the basic building blocks of nature. Atoms or pieces of atoms combine in various ways to create all the material in the universe.

The simplest atomic combination is the element hydrogen, which is a very light gas at room temperature. (See Figure 12–1.) It has one proton at the center of the atom and one electron whirling around the proton, something like a planet whirling around the sun. Hydrogen was one of the first elements created at the beginning of the universe. Other elements, i.e., helium, oxygen, copper, gold, lead—more than one hundred altogether—were formed as additional protons, neutrons, electrons, and other parts were added to the original hydrogen atoms (see Figure 12-2).

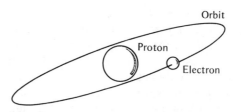

FIGURE 12-1. Hydrogen, the simplest atom.

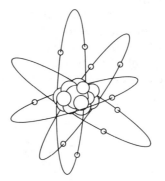

FIGURE 12-2. A more complex atom.

The atomic parts "stick" together to create atoms because nature strives for balance everywhere, from the very small to the very large. In atoms, the balance we are interested in is between protons and electrons. Protons are said to have a positive charge, like the south end of a magnet. Electrons are said to have a negative charge, like the north end of a magnet. In order for an atom to be balanced electrically, it must have an equal number of electrons and protons.

When an element is created, protons are packed together in the center, or nucleus, of the atoms. Electrons are attracted in equal numbers to orbits surrounding the nucleus. Because the electrons have the same charge and tend to repel one another, each orbit can only contain so many electrons, two in the first orbit, six in the second, and so on.

The last or outermost orbit is the most important in our understanding of electricity. The number of electrons in this orbit determines whether an element will be a conductor, non-conductor, or semi-conductor of electricity. It is in this orbit that electrical flow takes place.

ELECTRICAL FLOW

To see how electrical flow works, we will take an imaginary trip into the atomic structure of a copper wire. Copper, a conductor of electricity, has 32 electrons. That gives enough electrons to fill all but the last orbit, which has only one electron. Because these single electrons are so far from the nucleus and because they aren't bound in a fixed pattern with other electrons, they tend to drift from atom to atom. From our imaginary vantage point, the outer orbit electrons might resemble a swarm of fireflies, wandering here and there in response to random influences.

As long as the wire isn't hooked into a live electrical circuit, the pattern of electron movement remains random. An atom in one place loses an electron. Another electron, from somewhere else, takes its place, attracted by the extra positive charge exerted by the atom that has lost its electron. Since the movements are random, the sum total of all the electrical charges on all the atoms remains balanced (see Figure 12-3).

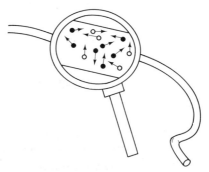

FIGURE 12-3.   Random electron movement.

However, what happens if the ends of the wire are attached to the positive and negative terminals of a battery? The negative pole contains an excess of electrons and the positive pole has fewer electrons than it needs to be balanced. The excess electrons at the negative pole repel the electrons in the wire, causing them to flow toward the other end. At the same time, the electrons at the other end of the wire are attracted to the empty orbits at the battery's positive pole. And almost simultaneously with that, the excess electrons from the battery's positive pole rush into the wire to fill the newly vacated orbits (see Figure 12–4).

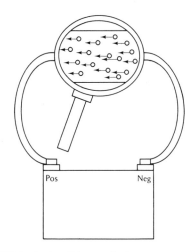

FIGURE 12–4.   Directed electron movement.

It's like a game of tag played at the speed of light, where nobody (or no electron) ever catches anybody. The chemical activity inside the battery constantly creates an imbalance between the two poles. So long as the battery remains active and the circuit is complete, the electrons will chase each other from empty orbit to empty orbit, through the battery and through the wire. If the battery weakens, the current slows down. If the wire is cut, then the flow stops altogether. However, the battery still pulls electrons away from the end of the wire connected to the positive pole and builds up electrons at the end connected to the negative pole

Electrical flow takes place wherever there is an overabundance of electrons in one place, an underabundance in another place, and an uninterrupted path between the two. The potential for electrical flow exists whenever there is an imbalance but no complete path for current to flow. For instance, even though no current flows when the ignition switch in a car is off, the potential for current flow still exists because there is an electrical imbalance on either side of the switch.

Current flow can be temporary or sustained. The sudden discharge of electricity represented by a lightning bolt is obviously temporary. An electrical imbalance is created between a cloud and the ground. When the potential becomes great enough, the excess electrons are able to create a path through the atoms in the air.

Most sustained current flow is the result of some man-made device: a battery, alternator, generator, etc. They convert either chemical or mechanical energy into electrical potential. That is, they create regions of electrical imbalance, which, when connected to a complete circuit, will cause electrical flow.

CONDUCTORS, NON-CONDUCTORS, AND SEMI-CONDUCTORS

The number of electrons in the outer orbit of an atom determines how readily substances containing the atom will allow electrical flow.

Conductors

Substances containing atoms with one to three outer orbit electrons are usually considered to be good conductors. The electrons in their outer orbits are loosely held and easily put into motion. Copper, gold, iron, silver, and water are examples.

Non-conductors

Materials containing atoms with five or more outer electrons are poor conductors under normal circumstances. Their electrons are tightly held in a pattern with other electrons and are not easily dislodged. Non-conducting materials like rubber, glass, and many plastics are used as insulators to channel and direct electrical flow—making sure it goes only where it is supposed to go.

Semi-conductors

Semi-conductors are very stable elements. Their atoms, containing four outer electrons each, join together in rigid patterns that are not easily disturbed and that do not readily support electron flow. However, this holds true only so long as the element remains in its pure state. If small amounts of impurities are added, the semi-conductor becomes a conductor. Depending on the kind of impurity added, current flow will depend primarily on the presence of positively charged "holes" (actually vacant spaces in outer orbits) or on free electrons floating through the semi-conductor (see Figure 12–5).

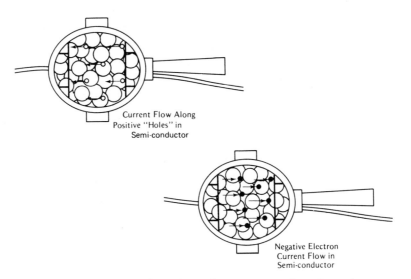

Current Flow Along
Positive "Holes" in
Semi-conductor

Negative Electron
Current Flow in
Semi-conductor

FIGURE 12-5.   Current flow through two types of semi-conductors.

In automotive applications, semi-conductors are used primarily to make diodes and transistors. Diodes are formed when two different kinds of "impure" semi-conductor chips are sandwiched together. Diodes, which are found in alternators, only allow current flow in one direction. They are like one-way current valves (see Figure 12-6). Transistors are formed by sandwiching three semi-conductors together. In most automotive applications, the two outside layers are made from one type of semi-conductor and the middle layer from the other type.

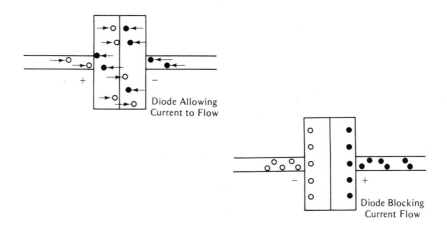

Diode Allowing
Current to Flow

Diode Blocking
Current Flow

FIGURE 12-6.   Diodes as one-way valves.

The middle layer is often connected to a low power control circuit. The outer layers are connected to a higher powered main circuit. Owing to the way the transistor functions, low level fluctuations in the control circuit will cause corresponding (but magnified) changes in the main circuit, including shutting it off if the control circuit is shut off. In this way, the transistor acts as a kind of amplifier (see Figure 12–7).

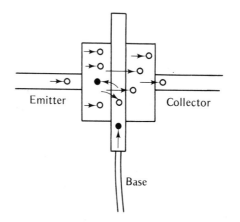

FIGURE 12–7.   Transistors.

EFFECTS OF ELECTRON FLOW

Electron flow alone doesn't have much practical application. The effects of electron movement are what we put to everyday use. For instance, electron flow jostles all the electrons in a circuit. These bouncing electrons cause atoms to vibrate, which results in heat. All current flow involves some degree of heating. If uncontrolled, electrical heating can be dangerous. A wire can melt its insulation and cause a fire. However, by using the right kinds of materials in properly designed circuits, the heat can be put to practical use—in a heater, an oven, or a diesel glow plug.

Another related effect of the atomic disturbance is light. Every time an electron bounces back and forth between the orbits in an atom, light flashes from the atom. As Thomas Edison proved, by using the right kinds of materials in the right circumstances, practical electric lights are possible (see Figure 12–8).

Besides having an effect on the atoms that make up a conductor, electron flow also causes a change in the surrounding space. When current flows through a wire, revolving rings of magnetic force are created around the wire. These lines of magnetic force, as noted in the second half of this chapter, have many applications as in electric motors, electric coils, etc.

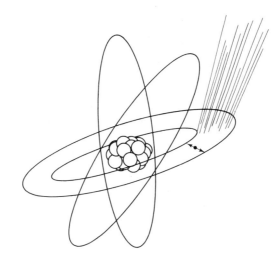

FIGURE 12-8.   Light emission from an atom.

ELECTRICAL MEASUREMENT

We've seen now what electrical flow is, why and how it works, and some of its effects. The last item to be considered before going on to magnetism is electrical measurement.

Volts

Voltage is a measure of electrical pressure. It is an indication of the pressure exerted on each electron in a circuit by a source of electrical potential. As the following example shows, it is *not* a measure of the total electrical force felt by all the electrons in a circuit.

To see what this means, imagine you have two 12-volt batteries, a large one that is used in a full-size pickup truck, and a small one that is used in a compact automobile. Both have the potential for exerting 12 volts of force or pressure on each electron in a circuit. In other words, the electrical imbalance existing at the positive and negative poles of both batteries will have the same disruptive effect on a given electron in a given atom. However, the larger battery has the potential for disturbing more electrons and more atoms.

Amps

Amperage is a measure of total electrical flow. It is an indication of the number of electrons flowing past a given point in a circuit. Amperage (or the number of electrons in motion) depends on electrical pressure (voltage) and on electrical resistance (ohms, discussed next).

Ohms

An ohm is a unit of electrical resistance. It is used to express the varying ability of different materials to support current flow. Some factors affecting resistance are: (1) the size of the conductor—just as a larger water pipe will let more water flow than a small pipe, a large conductor will let more electrons flow; (2) the length of the conductor—the longer the conductor, the greater the resistance; and (3) the nature of the conductor—as we've already seen, some materials support current flow better than others.

OHM'S LAW

As you might imagine, ohms, amps, and volts are related quantities. If one quantity changes, at least one of the others must also change. For instance, in a given circuit:

1. If voltage changes (goes up or down) and the ohms of resistance stay the same, then the amperage must also change (go up or down).
2. If the ohms go up (as a result of a frayed, or otherwise damaged wire) and the voltage at the source doesn't change, the amperage must go down. Not as many electrons can flow.
3. Conversely, if the ohms of resistance drop, the amperage must increase.

These relations are expressed in a statement called Ohm's Law. That law and associated formulas are given in Figure 12-9.

WHAT IS MAGNETISM?

Magnetism is a mysterious force operating equally well through air, empty space, or solid matter. No one knows exactly how it works. Useful theories have been devised that help predict how magnetism be-

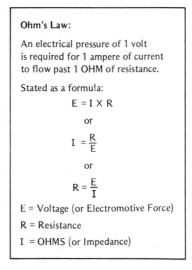

Ohm's Law:

An electrical pressure of 1 volt is required for 1 ampere of current to flow past 1 OHM of resistance.

Stated as a formula:

$$E = I \times R$$

or

$$I = \frac{R}{E}$$

or

$$R = \frac{E}{I}$$

E = Voltage (or Electromotive Force)

R = Resistance

I = OHMS (or Impedance)

FIGURE 12-9. Ohm's law.

haves and how it can be put to practical application. We will examine some of these theories because they are needed to help understand automotive electrical operation.

Like electricity, magnetism can still be neither felt, seen, nor touched. It is more a description of an effect than of a thing.

WHAT IS A MAGNET?

A magnet itself is easier to describe than magnetism. Magnets are objects with the power to attract or repel other magnets and to attract iron and certain materials made from iron. Magnets are directional; that is, they have distinct ends, or poles. If a straight or bar magnet is allowed to hang freely in a horizontal position, one end (called the north pole) will always point in a northerly direction and the other end (called the south pole) will always point toward the south. No matter how a magnet is cut, shaped or altered, it (or its pieces) will have two magnetically different poles.

THEORY OF MAGNETIC OPERATION

Scientists believe that magnetic properties originate at the atomic level. This is a slightly altered version of the explanation most often given.

Every atom in the universe is thought to spin on its axis, like the earth spins on its axis, or a toy top spins about its handle. The spinning action (along with electron movement) causes the atoms to be surrounded by lines of magnetic force. No one knows exactly what these lines of force are, or if they even exist in the usual physical sense. However, for our purposes, we can visualize them as being the paths taken by imaginary particles flying around and through atoms. The particles travel out one end of an atom's axis (its north pole), circle around the atom, re-enter at its south pole, then go through the atom to start the trip again (see Figure 12–10).

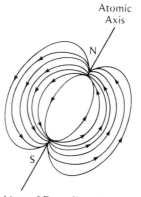

Atomic
Axis

Lines of Force (Imaginary
Particles) Passing Through
Atom

FIGURE 12–10. Magnetic force lines passing through a single atom.

In most materials, the atomic poles point in different directions. The lines of force surrounding adjacent atoms don't line up; the imaginary particles flying around neighboring atoms bump into one another as often as they travel along parallel paths.

However, something happens in magnetic materials. A substantial number of the atomic poles point in the same direction. This means that a substantial number of the imaginary particles travel in the same direction. Instead of expending their imaginary energy in headlong collisions, the nonexistent particles fly along parallel paths. They join forces to fly from atom to atom, building up enough momentum to actually escape, for awhile, the confines of the magnetic material (see Figure 12–11).

Random Atomic Alignment

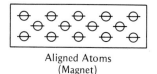

Aligned Atoms
(Magnet)

FIGURE 12-11. Aligned vs. unaligned atoms.

The action duplicates on a large scale what takes place on a small scale within atoms. Imaginary particles fly along parallel paths within a magnet, building up enough power and energy to leave the magnet at its north pole. Then the particles circle back to re-enter the magnet at its south pole and start the trip again. The paths taken by the particles represent lines of magnetic force—which extend in three dimensions on all sides of the magnet (see Figure 12-12).

The flight of the imaginary particles helps to explain how magnetic materials attract and repel each other. When the opposite poles of two magnets are placed together, the imaginary particles travel in the same direction, just as they do within the atomic structure of the two magnets. Consequently, the particles join forces, pulling the two magnets together. In effect, this creates a single magnet with double the power of the individual magnets.

However, if the like poles of two magnets are put in close proximity, the particles move in opposite directions. As a result, they bump together to push the magnets apart (see Figure 12-13).

Magnets are attracted to iron objects when the lines of force from the magnet penetrate into the atomic structure of the iron material. The iron atoms tend to line up with the lines of force, causing the imaginary particles to fly along with same paths. As the particles join forces, they tend to pull the two objects together.

These magnetic flight paths (whatever they represent) can actually be seen by sprinkling iron filings on a piece of paper that has been placed over a magnet. Tapping the paper lightly loosens the filings so they can follow their natural attraction to the lines of force. The pattern of filings represents the shape of that part of the three-dimensional magnetic field sliced through by the paper. Figure 12-14 pictures the magnetic fields surrounding horseshoe and bar magnets.

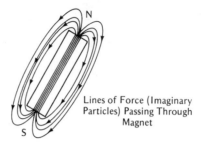

Lines of Force (Imaginary
Particles) Passing Through
Magnet

FIGURE 12-12.   Lines of force passing through a bar magnet.

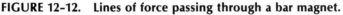

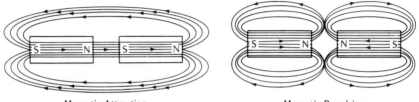

Magnetic Attraction   Magnetic Repulsion

FIGURE 12-13.   Attraction and repulsion between two magnets.

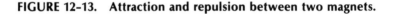

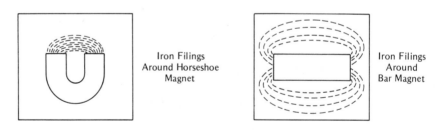

Iron Filings
Around Horseshoe
Magnet

Iron Filings
Around
Bar Magnet

FIGURE 12-14.   Iron filings around horseshoe and bar magnets.

KINDS OF MAGNETS

There are three basic kinds of magnets: permanent, temporary, and electromagnetic.

Permanent Magnets

Permanent magnets are often made from the mineral magnetite, which is a naturally occurring magnet. Its atoms are sufficiently aligned to produce a coherent magnetic field. Commonly called "lodestone," magnetite was discovered hundreds of years ago. Early mariners used it as a crude compass to help chart their course when travelling out of sight of land.

Temporary Magnets

Temporary magnets are created by placing certain iron or iron-based materials in the presence of strong magnetic fields. The lines of force from the magnetic field "pull" the iron atoms into alignment. Depending on the strength of the magnetic field and on the nature of the material, the atoms may remain in alignment after the magnetic field is removed. However, as the name implies, these kinds of magnets may not last very long. Any sort of shock can disturb the fragile atomic alignment; this causes the atoms to become jumbled again and no longer capable of producing a coherent magnetic field.

Electromagnets

Electromagnets depend on a peculiar property of electricity: whenever current flows through a conductor, the atoms in the conductor line up sufficiently to produce a coherent magnetic field.

If the conductor is a wire, the lines of force form concentric rings around the wire. In other words, the imaginary particles fly in circular paths around the wire. The direction in which the non-existent particles fly, clockwise or counterclockwise, depends on the direction in which the current is flowing (see Figure 12–15).

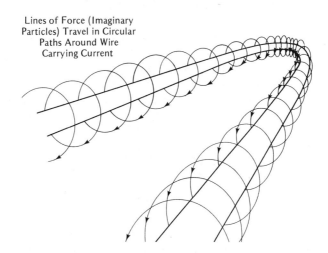

Lines of Force (Imaginary
Particles) Travel in Circular
Paths Around Wire
Carrying Current

FIGURE 12–15.   Force lines going around wire.

Electromagnets are created when current carrying wires are formed into a coil. This is how it works: Current flows in the same direction in adjacent loops of the coil. Consequently, the imaginary particles fly in the same direction in adjoining loops. As we've seen before, whenever these particles fly in the same direction, they join forces. So, the lines of force merge to circle all the loops, travelling around the outside of the coil, going back through the inside of the coil, then going around the outside again (see Figure 12–16).

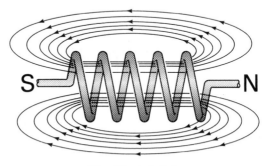

When Wire is Coiled,
Force Lines Merge to
Travel Around and Through
Entire Coil

FIGURE 12–16.   Wire formed in an electromagnetic coil.

An electromagnet is like a permanent magnet in many ways. It has a north pole (where the particles come out of the coil) and a south pole (where the particles return). An electromagnet can attract or repel other magnets. It can also attract iron.

However, there are still some important differences between the two. First of all, electromagnets can be turned on or off by turning the current on or off. The polarity of the electromagnet (which end is north and which is south) can be reversed by reversing the current flow. And the strength of the magnet can be varied by varying the current flowing through the coil or by increasing the number of loops in the coil.

AUTOMOTIVE APPLICATIONS OF MAGNETISM

We would not study magnetism if it did not have some practical use in automobiles. As it so happens, magnetically operated devices are found in the starting system and the charging system of diesel-powered vehicles and in the ignition system of gasoline-operated cars. The following paragraphs briefly introduce the operation of solenoids, relays, electric motors, and chargers, the four principal applications of magnetism in diesel operation. The remaining parts of this section cover these devices in more detail.

Solenoids

Solenoids are electromagnetic coils used as in mechanical actuating devices, attracting or repelling parts of a mechanical linkage. A solenoid coil is usually hollow. Often called a "sucking coil," it has the power to attract objects inside the coil itself—or to repel objects from the coil (see Figure 12-17).

Relays

A relay is an electromagnetic coil used (most often) in an electrical switch. Usually the relay coil is wrapped around an iron core. That is because metals with a high degree of magnetic "permeability" possess the power to concentrate lines of magnetic force, thereby increasing the strength of the magnet in the region of the core. It's as if the imaginary particles prefer to pass through certain materials. So, though the total number of force lines is not increased, the concentration in certain regions is increased (see Figure 12-18).

Electric Motors

Electric motors contain two sets of magnets, the field and the rotor. The field may be an electromagnet or a permanent magnet. The rotor is always an electromagnet.

Figure 12-19 illustrates a very simple electric motor. The field is a horseshoe magnet and the rotor is a single loop of wire placed between the ends of the horseshoe magnet. The rotor is connected by a switch (called a split ring commutator) to the power source.

In terms of the imaginary particles, this is how the motor works: Particles travel from the north to the south pole of the field magnet. At the same time, particles circle the rotor loop, going around one way as the current flows down one side the loop, and revolving the other way as the current flows back.

The end view of the motor shown in Figure 12-20 shows how the particles interact to turn the rotor. The top part of the rotor is pulled to the left because the particles from the field and the rotor move in the same direction on the left side of the loop and on the opposite direction on the right side of the loop. The particles on the left try to join up—which pulls the top of the loop to the left. At the same time the particles on the right push apart—which pushes the top to the right.

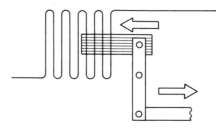

Solenoid . . . Coil Used
to Actuate Mechanical
Linkage

FIGURE 12-17.   A solenoid.

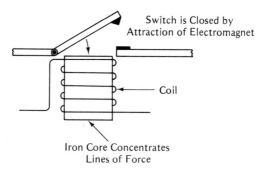

Switch is Closed by
Attraction of Electromagnet

Coil

Iron Core Concentrates
Lines of Force

FIGURE 12-18.   A relay.

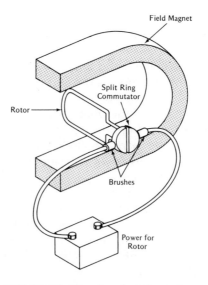

Field Magnet

Split Ring
Commutator

Rotor

Brushes

Power for
Rotor

FIGURE 12-19.   An electric motor.

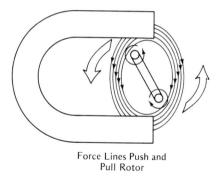

Force Lines Push and
Pull Rotor

FIGURE 12-20. An end view of a motor showing the reaction between field and rotor force lines.

A similar action takes place at the bottom of the loop. However, since the current (and hence the particles) move in the opposite direction, the pushing/pulling is to the right.

These forces move the rotor halfway around as noted in Figure 12-21. Momentum then takes the rotor on a little further. However, unless something happens, the forces acting on the rotor will cancel out at this point. They will work in opposite directions and the rotor will stop.

That is where the split ring commutator comes into play. As noted in Figure 12-22, it reverses the current flow through the loop. The two halves of the commutator swap connections with the power source. What was once the negative connection to the rotor becomes the positive lead and vice versa. As a result, the particles or lines of force circle the loop in opposite directions. Therefore, the forces acting on the rotor are reversed, causing it to continue on around. This cycle is repeated once every revolution of the rotor.

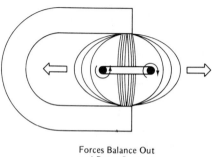

Forces Balance Out
and Rotor Stops at
Halfway Point

FIGURE 12-21. A rotor after turning halfway around.

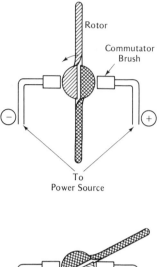

Rotor

Commutator

Brush

−

+

To
Power Source

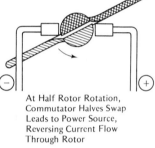

−

+

At Half Rotor Rotation,
Commutator Halves Swap
Leads to Power Source,
Reversing Current Flow
Through Rotor

FIGURE 12–22. A commutator reversing current flow.

Electric Generators (Induced Current Flow)

Theoretically, at least, most electric motors can be turned into electric generators. It's possible because of a peculiar property of conductors and magnetism called induced current flow.

Whenever a conductor and magnetic lines of force move in relation to one another, current flow is generated (or induced) in the conductor. For instance, if one side of a loop of wire is passed between the ends of a horseshoe magnet, current will flow through the wire. The same thing happens if the wire is held stationary and the ends of the magnet move (see Figure 12–23).

No one knows exactly how current flow is induced. The explanations that do exist are very complicated. For our purposes, we can simply say that the imaginary particles knock electrons lose from the outer orbits of atoms.

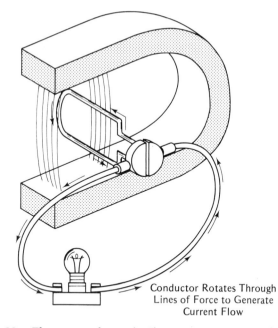

Conductor Rotates Through
Lines of Force to Generate
Current Flow

FIGURE 12-23.   The motor shown in Figure 12-22 converted to a generator.

The direction in which the electrons travel depends on the relative motion between the conductors and the lines of force. If the imaginary particles cut through the atomic structure of a conductor from one direction, the electrons will be knocked one way. If the particles strike in the other direction, the electrons will move that way.

So, the simple electric motor described before can be converted into a generator by turning the rotor between the ends of the horseshoe magnet. As the two sides of the loop cut across the magnetic field, current is induced. It flows through the ends of the loop, through the two halves of the commutator and on into the attached circuit.

However, something happens when the rotor reaches the midpoint of its travel, as noted in Figure 12-24. The two sides of the rotor loop start to cut back through the lines of force in an opposite direction. Consequently, the current will flow in an opposite direction.

In our simple motor/generator, the commutator can be used to keep the current flowing in the same direction. As the current reverses in the rotor, the two parts of the commutator swap sides of the circuit. The commutator half that was feeding one half of the circuit now feeds the other half and vice versa. Therefore, as the current flow in the rotor is reversed, the leads from the rotor to the circuit are also reversed. Consequently, the flow throughout the circuit remains the same.

That is how mechanical force is converted into electrical potential. This potential can be used to operate other devices in the circuit or to recharge the power source noted before, or, in automotive operation, to do both.

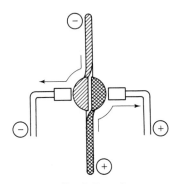

Current Flow in Rotor Reverses
at Half Rotation, HOWEVER . . .

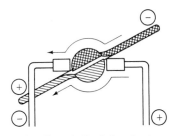

Flow Through Circuit Remains the
Same Because Commutator Swaps
Connections to Circuit at the Same Time

FIGURE 12-24.   The action of a commutator converting AC to DC.

CHAPTER 13

Diesel Electrical Circuits

GENERAL

Whenever a conductor moves in relation to magnetic lines of force or is connected to a source of electrical potential, electrons flow through the conductor. The endless path along which the electrons flow is called an electrical circuit. This chapter examines the basics of electrical circuitry and then looks at some circuits that are unique to diesel operation.

SERIES AND PARALLEL CIRCUITS

All circuits fall into one (or a combination) of two categories: series and parallel.

A series circuit is like a one-way road with no side or connecting roads. All electrons in motion in a series circuit must travel past every part and through every device in the circuit. All points of resistance must be overcome by every electron; any break in the circuit will stop current flow in the entire circuit. Figure 13-1 pictures a simple series circuit.

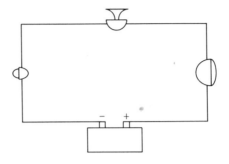

FIGURE 13-1.   A simple series circuit.

Parallel circuits are akin to a branching network of roads. Electrons may move along any one or all of the branches. A break in one part of the circuit does not necessarily interrupt the current flow to the other parts. By the same token, changes in resistance (either high or low) in one branch do affect current flow to the other branches. Figure 13-2 pictures a simple parallel circuit.

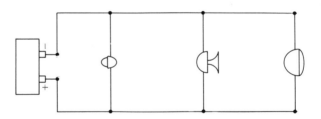

FIGURE 13-2.   A simple parallel circuit.

ELECTRICAL SCHEMATICS

The schematics, or wiring diagrams, used to picture electrical circuits are like roadmaps showing the main features, junctions, and points of interest along electrical pathways. Just as a roadmap lets a driver plan a trip between two geographical points, a schematic lets a mechanic chart a diagnostic course between the elements in an electrical circuit.

The items pictured on electrical diagrams include switches, batteries, alternators, fuses, solenoids, motors—virtually anything that might be found in a circuit. In some cases an item might be pictured literally, that is, it might appear on the diagram as it actually looks.

However, in most cases, circuit components are shown symbolically by characters or figures. Figure 13-3 pictures some common electrical symbols.

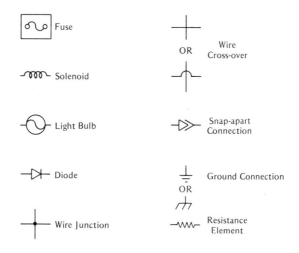

FIGURE 13-3.   Some common electrical symbols.

Electrical schematics often appear very complicated. One secret to using them is to concentrate on the problem at hand—since the only reason to go to a wiring diagram in the first place is to diagnose an electrical malfunction. In other words, after locating the likely trouble spots on the schematic and finding the segments of the circuit that connect these elements, ignore everything else.

CIRCUIT FEATURES AND CONDITIONS

However, even before applying this rule, you need to know certain facts about electrical circuits. You wouldn't attempt to use a roadmap without knowing the difference between various kinds of roads, intersections, etc. You also can't expect to use a wiring diagram without knowing something about various circuit features and conditions.

Ground Return Circuits

In the simple circuits illustrated in the preceding, the current flows along wires to and from the power source. However, in automotive applications, the conductive parts of the frame, engine, and body are used for almost half of the circuitry. The negative terminal of the battery is connected to the frame and/or engine block. Electrons from the battery flow into any wire or device connected to the frame, body, or block. Current flows back to the positive terminal of the battery along wires and cables (see Figure 13-4).

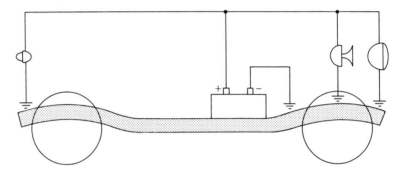

FIGURE 13-4.　Ground returns in an automotive circuit.

In effect, the chassis, body, and engine become an extension of the negative pole of the battery, or one side of a source of electrical potential. Attach a wire to any conductive part of the vehicle; connect the other end to the battery's positive terminal and current will flow.

Open Versus Closed Circuits

Current only flows if there is a closed path between the ground point and the positive terminal of the battery. So long as there is an intentional break (like an open switch) or an unintentional break (like a severed wire), electrons will not flow. Only the potential for flow will exist—represented on the negative side of the opening by an excess of electrons and on the positive side by a reduction of electrons (see Figure 13-5).

Hierarchy of Switches

Therefore, every open switch in an automobile can be said to straddle two sides of an electrical potential. However, this potential does not exist at all switches at all times. For example, in most cars, even if you turn on the switch to the electric wipers, the wipers will not operate. That is because automotive electrical circuits contain a hierarchy of switches wired in series with one another—in other words, switches that control the current flow to other switches. The master control for most automotive circuits is the ignition switch. Until it is closed, the other switches will not operate (see Figure 13-6).

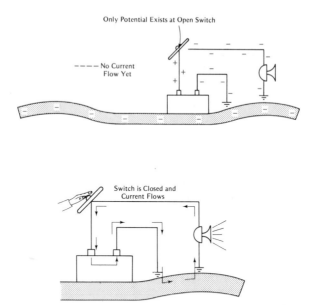

FIGURE 13-5. Electrical potential vs. electrical flow at open and closed switches.

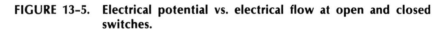
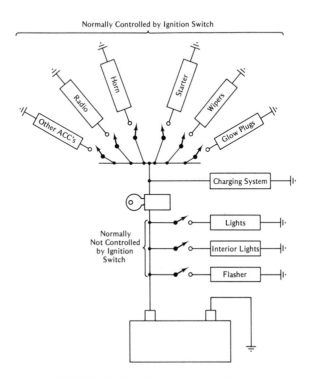

FIGURE 13-6. Hierarchy of switches.

Series/Parallel Circuit Arrangements

It is interesting to note that after a switch is closed, current does not usually flow through just one wire, but several. For instance, when the headlight switch is closed, current flows to the two headlights, two taillights, and probably some side lights as well. These components are wired in parallel so that if one fails, current can still flow to the others. The same is true for many other automotive circuits. However, some components, because of the functions they perform, must be wired in series. Switches are one example. Any switch must exist in a series relationship with the element it controls. Otherwise, turning the switch on or off would have no effect. This holds true whether the controlled element is another switch or an electrically operated device. Fuses and breakers are another example of series wired components. As we'll see in the next paragraphs, they are designed to protect groups of circuits by deliberately breaking the circuits if current flow exceeds certain limits; this dictates a series relationship (see Figure 13-7).

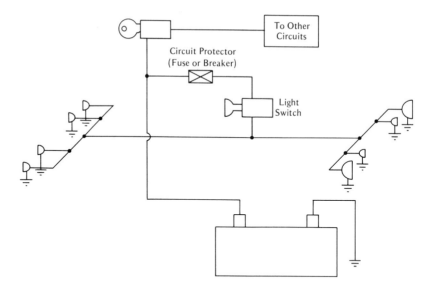

FIGURE 13-7.   Series/parallel circuit arrangements.

CIRCUIT ABNORMALITIES

So far, we've examined various aspects of normal circuits. Next, we'll look at some abnormal or problem conditions.

Short Circuits

A short circuit occurs when two wires come into accidental contact. A short represents a junction where none was intended. Because current flows to areas of least resistance, it is possible that all or a substantial portion of the current flowing in both of the shorted wires will be shunted through the one with the lowest resistance. As a result of the added current load, that wire may overheat, possibly enough to cause a fire. By the same token, the wire with reduced current flow will not be able to supply the devices to which it is connected.

Grounded Circuit

A grounded circuit is like a short circuit with one important exception. Instead of two wires coming into accidental contact, a wire touches ground where it shouldn't (e.g., the wire touches part of the chassis, body, or engine). Since the ground circuit offers very little resistance to current flow, the battery tries to push more and more electrons through the ground contact point. It is like opening a floodgate into the wire. As a result, the wire overheats and the battery becomes drained (depending on the location of the ground).

Open Circuit

An open circuit refers to an unintentional break due to a damaged connection, a severed wire, etc.

High Resistance Circuit

Excessive resistance occurs when a wire or connection is damaged or corroded enough to restrict current flow but not block it altogether.

CIRCUIT PROTECTORS

Any of the conditions described above can cause problems. However, shorted and grounded circuits are potentially the most dangerous. By shunting excess current through part of a circuit, they can result in overheating, which can cause fires.

Fuses and circuit breakers, as noted previously, are designed to detect excessive current flow.

Fuses

Fuses are metal strips mounted on ceramic bodies or in glass encased shells. All the current for the circuit protected by the fuse must pass through the metal strip. If the current exceeds the design limits of the circuit, the strip overheats and melts apart, thereby blocking the flow to the remainder of the circuit.

The size or rating of a fuse is usually stamped somewhere on its body. The rating indicates the maximum current that should flow through a given circuit. For instance, a 5-amp fuse is designed to fail if the current exceeds that amount long enough for the metal strip to melt. Therefore, it's always important to match the right fuse to the right circuit.

Breakers

Breakers, like fuses, are designed-in weak places in a circuit. The principal element in most breakers is a bi-metal spring. It consists of two different kinds of metal sandwiched together. When heated, these metals expand and contract at different rates. As a result, the bi-metal spring curls up when it is heated, and uncurls when it cools.

This feature is used in breakers by passing current through the bi-metal spring the same as it is passed through the metal strip in a fuse. One end of the spring acts like a switch, capable of opening and closing the circuit. If the current exceeds the design limits, the spring curls away from the switch contact and opens the circuit.

Some breakers are designed to permanently open the circuit if the current limit is exceeded. They must be reset before operation is resumed. Other breakers allow current to flow on an intermittent basis, e.g., after the bi-metal spring cools off, it closes the switch, then heats up to open the switch, then cools to close it again and so on.

UNIQUE DIESEL CIRCUITRY

The electrical circuits in diesel automobiles and small trucks are similar to those used in gasoline vehicles. The following paragraphs note the principal differences.

Starting Loads

Because diesels operate at higher compression ratios, the engines are more difficult to crank. Therefore, diesel starter motors, although similar in design to their gasoline engine counterparts, are often larger and more powerful. One manufacturer, GM, uses two 12-volt batteries, wired in parallel to provide additional starting power.

Glow Plugs

These devices, sometimes called preheater plugs, are screwed into the cylinder head much like spark plugs. Glow plugs are small resistance heaters that provide the necessary increase in temperature to get the engine started. After that, the diesel combustion process sustains itself and the glow plugs are shut off, not only because they aren't needed, but because they consume a great deal of electrical power. Figure 13-8 shows a Peugeot glow plug.

Figure 13-9 pictures the wiring diagram of the Nissan diesel glow plug circuit. Note that all the glow plugs are connected in parallel and that a glow plug pilot light is used to let the operator know when the glow plugs are in operation.

Figure 13-10 shows the glow plug electronic control circuit used in 1980 GM diesels. This sytem employs 6-volt glow plus operating on the dual battery 12-volt system for faster warmup. An electronic control module is also provided. It monitors temperature data sent by a thermal probe (thermistor) located at the front of the intake manifold. Should a malfunction occur, the electronic control module will shut down the glow plug system and flash a message on a signal panel located on the car dash.

FIGURE 13-8.   A glow plug (*Courtesy of Peugeot*).

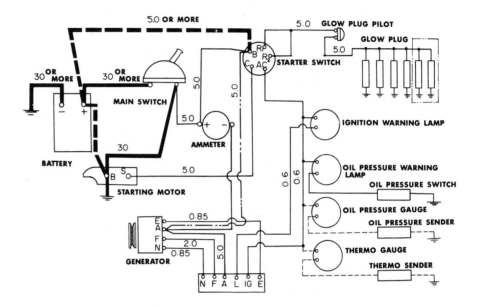

(1) - - - - - - - -   indicates that an electrical oil pressure gauge and water temperature gauge are installed.

(2) ▬▬▬▬▬▬▬   indicates that a main switch is not used.

(3) — - — - — -   indicates that an ammeter is not used.

(4) There are two type of oil pressure gauges and water temperature gauges available Electric type and Bourdon Tube type.

(5) The values given on the wiring diagram are reference values (mm) of the sectiona areas of the wiring used.

FIGURE 13-9.   A Nissan glow plug circuit (*Courtesy of Nissan Diesel Motors Ltd./Marubeni America Corporation*).

SECOND TYPE DIESEL GLOW PLUG ELECTRICAL CONTROL

	1	2	3	4	5 ENGINE RUNNING	6

1
IGN. SWITCH
-"OFF"-

WAIT LAMP - OFF
START LAMP - OFF
GLOW PLUGS - OFF

2
IGN. SWITCH
-"RUN"-

WAIT LAMP - ON
START LAMP - OFF
GLOW PLUGS - ON

3
IGN. SWITCH
-"RUN"-

WAIT LAMP - OFF
START LAMP - OFF
GLOW PLUGS - ON
↓
SEE NOTE 1

4
IGN. SWITCH
-"START"-

WAIT LAMP - OFF
START LAMP - ON
GLOW PLUGS - ON

5
IGN. SWITCH
-"RUN"-

WAIT LAMP - OFF
START LAMP - ON
GLOW PLUGS - ON
↓
SEE NOTE 2

6
IGN. SWITCH
-"RUN"-

WAIT LAMP - OFF
START LAMP - OFF
GLOW PLUGS - OFF

NOTE 1: If the ignition is left in the "Run" position without starting the engine, the glow plugs will continue to pulse on/off until batteries run down. (About 4 hours when coolant switch is open.)

NOTE 3: Do not manually energize or by-pass the glow plug relay as glow plugs will be damaged instantly.

NOTE 2: Glow plugs will pulse on/off for about 30 seconds after engine starts. Then turn off and remain off as long as coolant temperature is above about 120°F (49°C).

NOTE 4: Diodes prevent glow plug operation when the engine is warm (above 120°) the engine is not running and key is in RUN.

IMPORTANT: Do Not use more than a 2–3 candle power test light when making circuit checks.

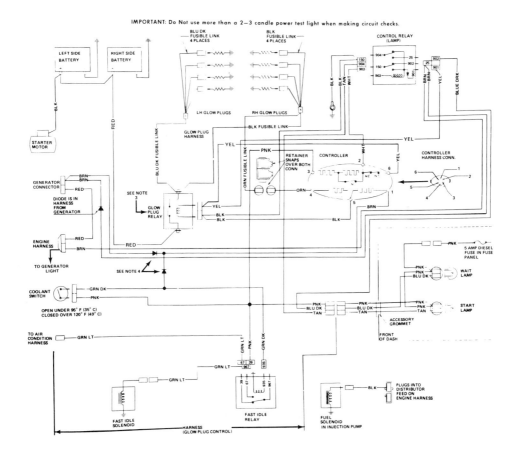

FIGURE 13–10. A General Motors glow plug circuit (*Courtesy of Oldsmobile Division, General Motors Corporation*).

CHAPTER 14

Batteries Used in Diesels

INTRODUCTION

Because of heavier starting loads, the batteries used in diesel cars and light trucks are likely to be more powerful than those found in gasoline operated vehicles. Some diesel manufacturers may even use two 12-volt batteries wired in parallel for added power. However, except for these differences, the batteries found in diesels are essentially the same as batteries in cars with gasoline engines. This chapter reviews some basic principles of battery operation and construction.

BASIC OPERATION

In its simplest form, a battery contains: (1) a pair of dissimilar metal plates, (2) positive and negative terminals attached to these plates, (3) a chemically active liquid called electrolyte in which the plates are immersed and, (4) a container for the plates and the electrolyte (see Figure 14-1).

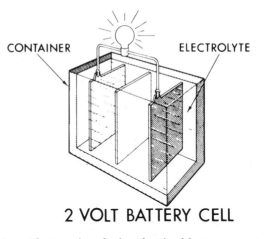

CONTAINER   ELECTROLYTE

2 VOLT BATTERY CELL

FIGURE 14-1.   **A wet battery is a device that is able to convert chemical energy to electrical energy. The chemical action can be re-
versed to recharge the battery. (*Courtesy of Ford Motor
Company of Canada, Ltd.*).**

Electrons are moved because of the chemical reaction between
the electrolyte and the two kinds of plates. Very roughly speaking, the
electrolyte allows electrons to move away from one kind of plate and
build up on another. As a result, one kind of plate and its attached terminal will have more than the normal number of electrons. The other
plate and its terminal will have fewer electrons. In the plate with fewer
electrons, the positive protons will predominate, giving that plate and
its terminal a positive charge. The plate and terminal with more electrons will have a negative charge.

These same basic features can be expanded in a number of ways
to create a more complex battery. For instance, extra pairs of positive
and negative plates can be added, in a kind of alternating "sandwich"
(so long as the layers don't touch). These extra plates will not increase
the voltage or "push" given to any single electron released by any particular plate. However, the capacity of the battery is increased by increasing the total number of electrons put into motion.

The simplified battery can also be combined with other batteries.
Its positive and negative terminals can be joined to the opposite terminals of a second battery, and that battery's terminals to the opposite
terminals of a third battery and so on. Joining batteries this way, (pos.
to neg., to pos. to neg., to pos., and so on) will add to the total electrical
"push" or voltage. For instance, if 6 two-volt batteries are joined together, the total pressure will be (6×2) 12 volts (see Figure 14-2).

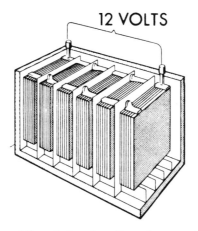

12 VOLTS

FIGURE 14-2.   Assembling six 2-volt cells and connecting them in series inside a single battery case creates a 12-volt battery. (*Courtesy of Ford Motor Company of Canada, Ltd.*).

When several small batteries are combined in one battery case, the individual "sub" batteries are called "cells." So, in a way of speaking, a six-cell twelve-volt automotive battery is made up of six smaller batteries.

AUTOMOTIVE BATTERY CONSTRUCTION

Plates

Each cell in modern automotive batteries contains two groups of plates, a positive plate group and a negative plate group. The negative plates are made of porous, electrically conductive sponge lead. The positive plates are also porous lead that has been coated with lead peroxide paste. The plates are usually cast in a honeycomb-like grid. The horizontal and vertical ribs of the grid are made from a lead antimony alloy to give strength to the otherwise weak sponge construction.

Separators

If the alternating pairs of positive and negative plates touch, there will be no electron flow. In order to keep the plates apart, two kinds of separators are used. One kind is a non-conductive, porous sheet located between each of the positive and negative plates. The porous design allows electrolyte to flow between the plates. The non-conductive nature of the material prevents the separators from entering into the chemical reactions. Most modern separators provide vertical passages or grooves so that particles broken loose from the spongy plate material can drop into the "catch area" at the bottom of the battery. This secondary, cleansing function of the separators greatly extends the life of the battery (see Figures 14–3 and 14–4).

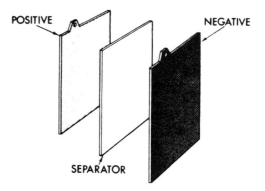

FIGURE 14–3. The basic components required in a battery cell (*Courtesy of Ford Motor Company of Canada, Ltd.*).

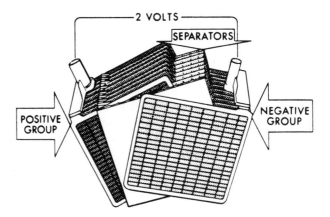

FIGURE 14–4. Automotive battery cells have a number of positive and negative plates insulated from each other by separator plates. Increasing the number of plates increases the surface area exposed to chemical action, thereby increasing capacity. Regardless of the number of plates per cell, each cell is capable of producing only approximately 2 volts. (*Courtesy of Ford Motor Company of Canada, Ltd.*).

Connector Strap Separators

The other kind of separator is the plate connector strap. There is one connector strap for all the positive plates in a cell and another for all the negative plates. These cast lead, electrically conductive straps have several functions. They help separate the positive from the negative plates, as well as join the negative and positive plates into two separate groups. Plus, the straps provide an electrical connection between the plates and the cell's terminals (the terminals, positive and negative, are where electrons enter and leave a cell). The strap connecting all the negative plates is joined to the negative terminal and the strap connecting all the positive plates is joined to the positive terminal. That way, an excess of electrons is built up not only on the negative plates, but also on the negative connector strap and the negative terminal. By the same token, there is a reduction of electrons on the positive plates, connector strap, and terminal (see Figure 14–5).

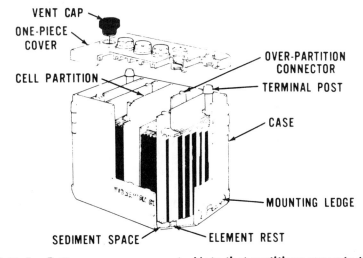

FIGURE 14–5.   Battery case components. Note that partitions separate the cells from each other. (*Courtesy of Ford Motor Company of Canada, Ltd.*).

Cell Connectors

In order for current to flow between cells, the terminals of one cell must be joined to the opposite terminals in the next cell (pos. to neg. to pos., and so on). One style of cell connector goes over the partition dividing the cells. In many newer batteries, the cell connector passes between the partitions. This reduces the distance the electrons have to travel and improves the battery's performance (see Figure 14–6).

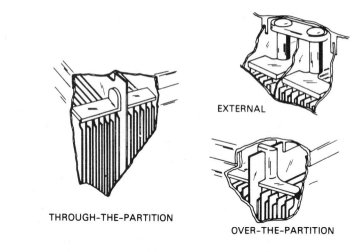

THROUGH-THE-PARTITION

EXTERNAL

OVER-THE-PARTITION

FIGURE 14-6.   Different types of cell connectors. Sealed top batteries use
the internal type of cell connectors. (*Courtesy of Ford Motor
Company of Canada, Ltd.*).

Battery Posts

The connector straps at both ends of the battery are joined to the battery's external positive and negative posts. These external posts or terminals may be identified in a variety of ways. The positive post may be
larger than the negative post, painted red, marked (+) on top or (pos.)
nearby. The negative terminal in turn may be smaller, painted green or
black, and marked (−) on top or (neg.) nearby. The terminals may be
on top of the battery, or, in some newer "energizer batteries," the terminals may be on the side of the battery case.

Battery Case

The battery case is the box or container which holds the plates, cells,
and electrolyte. Most modern battery boxes are made from a plastic,
such as polypropylene (see Figure 14-7).

ELECTROLYTE

Electron flow is a three-way chemical reaction between the electrolyte
and the two kinds of lead plates. In automotive batteries, the electrolyte is a solution of sulfuric acid and water, about 60% water and 40%

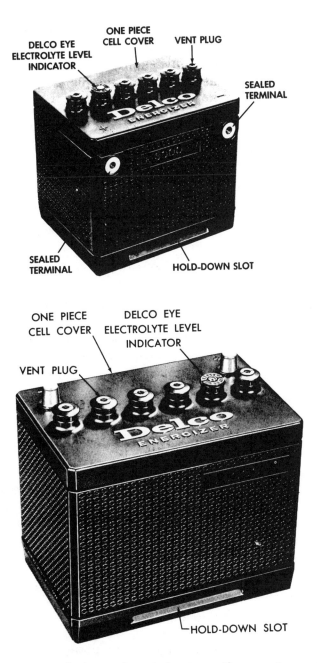

FIGURE 14-7.   Battery manufacturers do not always use the same types of battery terminals or terminal locations. (*Courtesy of General Motors Corporation*).

acid. The sulfuric acid may be said (greatly oversimplifying a complex reaction) to pull electrons from the lead peroxide, positive plates and allow electrons to build up on the negative sponge lead plates. This creates a "potential" for electron movement at the battery's two external terminals.

Because the electrolyte is made up of a sulfuric acid solution, it is dangerous. Spilled electrolyte from a battery will destroy paint and metal. It can also eat holes in clothing, burn exposed skin, and cause blindness if splashed in the eyes (see Figure 14–8).

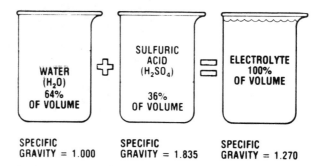

WATER (H_2O) 64% OF VOLUME

SULFURIC ACID (H_2SO_4) 36% OF VOLUME

ELECTROLYTE 100% OF VOLUME

SPECIFIC GRAVITY = 1.000   SPECIFIC GRAVITY = 1.835   SPECIFIC GRAVITY = 1.270

FIGURE 14–8. Composition and specific gravity of battery electrolyte. (*Courtesy of Chrysler Corporation*).

DISCHARGING

The term discharging simply means that the battery is being used, that it is connected to a complete electrical circuit through which electrons can flow.

Discharging causes the active ingredients in the electrolyte to be used up. The sulfuric acid dissolved in water breaks down into two kinds of particles, both of which have an electrical charge. The positive particles are called hydrogen ions and the negative particles, sulfate ions. Positive and negative charges are also created on the surface of the lead plates.

Because these charged particles are unstable, they try to combine with one another to become electrically balanced. Positive hydrogen ions from the acid combine with negative oxygen ions from the lead peroxide plates to form water. Negative sulfate ions from the electrolyte combine with the positive lead ions from both plates to form a chemically neutral coating on the plates called lead sulfate. The more the battery is discharged, the fewer active ingredients it contains until finally the electrolyte becomes almost pure water.

Also, as the battery discharges, the tiny holes and crevices in the spongy lead plates become coated with the neutral lead sulfate. This renders the plates ineffective (see Figure 14-9).

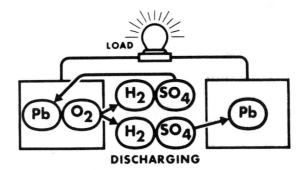

FIGURE 14-9. Chemical action inside battery as current is being used and battery is discharging. P_B is sponge lead, O_2 is oxygen; therefore, P_BO_2 is a lead oxide, H_2 is hydrogen, SO_4 is sulfate; therefore, H_2SO_4 is sulfuric acid.

CHARGING

Fortunately, the effects of discharging in lead acid batteries are reversible. If a flow of electrons is forced through the battery in a direction opposite the normal flow, the previous chemical reactions will reverse themselves. The sulfate ions from the plates and the hydrogen ions from the water will go back into the electrolyte. The battery will become active again.

When the engine is running, the battery is constantly being recharged by the alternator or generator. The car's charging system takes some of the fuel's energy and converts it into electrical power.

Depending on the electrical load and the operating conditions, the charging system will supply current at a pressure of 12 to 14.5 volts in a 12-volt system. If the battery is used in normal service and if the charging system is adjusted properly, a battery should last three to five years. However, if the battery is used improperly; for instance, if the lights are left on overnight, or if the charging system is not functioning, then the battery may discharge so much that it can't crank the car. In these cases, if the battery's plates aren't too badly coated, the battery can be reactivated by an external battery charger. It does the same job as the car's charging system and usually has two operating ranges: a fast charge and a slow charge. The fast charge supplies a great deal of current at high voltage pressures so the battery can be used sooner. The slow charge, which is generally safer and better, supplies a slow steady flow of current to gradually, thoroughly recharge the battery (see Figure 14–10).

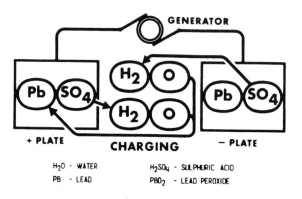

FIGURE 14–10.   **During the charging cycle, chemical action inside the battery is the reverse of that shown in Figure 14–9.**

BATTERY FAILURE

Batteries do not last forever. Following are several ways that batteries can fail:

1. Plate sulfation and electrolyte loss, as noted before, are natural results of the discharging process. Up to a point, this process is reversible by recharging the battery—either using the car's own charging system, or if the battery won't crank the car, an external charger. However, if the sulfation is too extensive, the battery must be junked (see Figure 14–11).

2. Battery cells also simply wear out. Repeated charging and discharging will, after a time, knock the active ingredients from the lead plates. These ingredients fall to the sediment chamber at the bottom of the case to form an inactive lead sulfate sludge. When enough sludge is at the bottom of the case, the battery no longer has the ingredients to develop any power and must be discarded.

3. Batteries can also be damaged by various mechanical means. Road shocks and vibrations can loosen the plates and cause "tree"-shaped cracks in the separators and case. On occasion,

a mechanic will pull an external post off trying to free a corroded battery cable.

4. Batteries are sometimes damaged by overcharging. This creates excess heat and causes gases to build up inside the cells. If the situation becomes bad enough, the plates, case, and separators may buckle (see Figure 14–12).

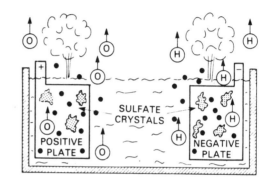

FIGURE 14–11. A battery becomes sulfated due to a discharged condition. Hardened sulfate crystals penetrate the pores of the plates. These crystals become insoluble. Prolonged charging at a low rate is required to restore it. Charging a sulfated battery at too high a rate can buckle the plates and destroy the battery. A battery that has been sulfated for too long cannot be restored.

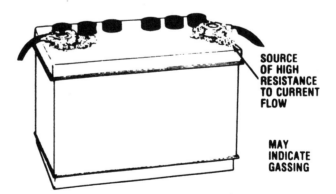

FIGURE 14–12. A charging rate that is too high will cause gassing and will result in sulfate deposits at the battery terminals. Electrolyte on the terminals can also cause the creation of deposits. (*Courtesy of Chrysler Corporation*).

5. An unused battery may damage itself simply because it is not being used. Unused batteries, unless they are emptied of electrolyte and stored dry, should be recharged from time to time. Otherwise enough spontaneous chemical activity will take place to sulfate the plates and use up the electrolyte.

BATTERY EFFICIENCY

Temperature

Atomic and molecular activity is dependent on temperature. When it is hot, electrons move faster; molecules bounce around quicker and chemical reactions proceed at a faster rate. The opposite holds true when it is colder. Everything slows down until a theoretical point called absolute zero is reached. At that temperature (or no temperature), there is no atomic activity.

These temperature/molecular-activity relationships hold true in the battery the same as they do anywhere in the universe. When it is cold, the battery's chemical reactions are sluggish and it develops less power. Then, when the temperature goes up, the reactions speed up and the battery develops more power. So, the battery is more efficient at higher temperatures.

Discharge Rate

Battery efficiency is also affected by the rate of discharge, in other words, by how much current the battery tries to move through a circuit connected to its terminals. When the resistance of the elements connected to the battery is low, it will try to push more electrons and its efficiency will be reduced. That's because the chemical reactions will all take place on the surface of the plates. The exchange of ions and electrons will not have time to penetrate into the interior of the spongy lead plates.

BATTERY RATINGS

Voltage Output and Variations

Each cell in a lead acid battery can produce about 2.13 volts of electrical pressure (usually rounded off to two volts). The cells can produce this voltage regardless of the size or number of the plates. However, the cell voltage rating is a maximum figure. In actual practice the voltage output may be less. Here are some of the factors that affect voltage output.

1. Temperature variations cause changes in voltage output. When the temperature is lower, the voltage is also lower.
2. The state of the electrolyte and the plates also affects voltage. When the electrolyte is weak and the plates sulfated, the voltage will drop.
3. Discharge rate likewise determines voltage output. When resistance to flow is reduced, the amperage will increase. And, according to Ohm's Law, the voltage will have to drop in order to balance the equation (relationship). (Of course, all these changes in voltage and amperage follow the relationships reflected in Ohm's Law.)

Amperage Output and Rating

Batteries are usually rated by their ability to produce current at certain conditions. Following are some of these rating tests:

The Twenty Hour Test. This popular way to rate batteries measures the amount of current a battery can deliver for 20 hours, without the cell voltage dropping below 1.75. For instance, a battery might be able to deliver 5 amps for 20 hours without the voltage dropping below 1.75. Usually the results of this test are given in battery advertising in terms of "AMP-HOUR" ratings. So, the battery just described would be rated at 100 amp hours because it could deliver 5 amps for 20 hours, or 100 amps for one hour.

The Battery Reserve Test. This measures the ability of a battery to supply current to the various accessories (lights, horn, radio, etc.) when the charging system is not operating. The results of this test are given in terms of the time (hours, minutes, etc.) it takes for the voltage to drop below 10.5 when the battery is delivering current at the rate of 25 amperes.

The Cold Tests. The above two tests are performed on batteries at 80°F. However, since the efficiency of the battery drops as the temperature goes down, it is also necessary to test the battery at cold temperatures. One cold test measures how long (hours, minutes, etc.) a battery at 0°F can deliver 300 amps before the cell pressure drops to one volt. Another cold test determines how many amperes a battery can deliver at 0°F and still maintain a pressure of 7.2 volts. This later figure was chosen by the SAE (Society of Automotive Engineers) as the minimum voltage required to provide adequate cranking speed for a large modern V-8 engine.

The Watt-Hour Test. This measures a battery's ability to perform work for a period of time. It is obtained by multiplying amp-hours times voltage. Some experts feel that rating a battery by its wattage potential is more accurate than the amp-hour method. They say the amp-hour method doesn't distinguish well enough between the power capabilities of 6-volt and 12-volt batteries. For instance, a 6-volt, 100-amp-hour battery has more watts of power (6×100=600) than a 12-volt, 45-amp-hour battery (12×45=540).

ELECTROLYTE SPECIFIC GRAVITY

As noted before, when the battery is discharged, the sulfuric acid in the electrolyte is used up. The active ingredients combine with the lead plates, causing the electrolyte to weigh less. The more the electrolyte weighs, the more acid it contains. The less it weighs, the less acid it will have.

The specific gravity test is a way to compare the weight of electrolyte with water. Water is said to have a specific gravity reading of 1.000. A battery filled with plain water would have virtually no electrical potential at its poles. So a battery whose electrolyte reads near "one" would have no power. All its acid would be used up and the electrolyte turned back to water.

However, a battery whose specific gravity is between 1.260 and 1.300 would be fully charged. In that range, the electrolyte contains all its active ingredients.

Specific gravity is usually tested by a device called a hydrometer. Each cell in a battery is checked by drawing a sample of electrolyte into the hydrometer. A float is suspended in the electrolyte sample and the specific gravity determined by checking the level of the float against a scale. The higher the float rides in the sample of electrolyte, the denser the liquid and the greater the specific gravity.

CHAPTER 15

Diesel Charging Systems

INTRODUCTION

This chapter briefly reviews the principles and operation of conventional, alternator based charging systems, the kind used in both diesel and gasoline powered automobiles and small trucks.

GENERAL

A storage battery, operating alone, can't supply an automobile's electrical needs for an extended period of time. The engine, the radio, lights, windshield wiper, lighter, and horn all require electrical power. To keep these systems functioning, automobiles must have a charging device.

The charger is operated by a V-belt connected to a pulley on the end of the crankshaft. In effect, the charger converts some of the crankshaft's mechanical energy (which was derived from the fuel's chemical energy) into electrical energy. This energy is used to charge the battery as well as supply the needs of the other electrical devices.

CHARGER-BATTERY RELATIONSHIP

Before getting into an explanation of alternators, it will be helpful to explore briefly the relationship between charging devices and batteries.

When a vehicle is first started, before the charger is rotating very fast, the battery supplies all the vehicle's electrical needs. Electrons move out of the battery's negative post, through the ground return circuit, to the vehicle's electrical load elements, and then back to the positive battery terminal to form a complete circuit. A regulating device of some kind prevents the battery from discharging back into the charger at these speeds.

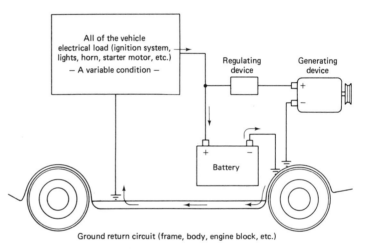

FIGURE 15-1.   When an engine is first started, the battery supplies all electrical needs (*Courtesy of Prentice-Hall, Inc.*).

After the battery has operated for a time, its ability to produce current is reduced. Also, as the engine speeds up, the charger, because of its basic design, is able to produce more voltage than the battery can deliver even when it is fully charged. So, electrons from the charger, as they move into the ground circuit, are able to move in a reverse direction through the battery's negative terminal. In other words, the charger's electromotive force (EMF) will be able to overcome the battery's counter electromotive force (CEMF). The charger supplies the vehicle's load requirements as well as pushes reverse current through the battery. The reverse current builds up and maintains the battery's charge level.

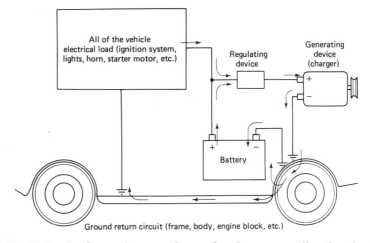

FIGURE 15–2. As the engine speeds up, the charger supplies the electrical load as well as recharging the battery (*Courtesy of Prentice-Hall, Inc.*).

If the vehicle's electrical load increases sufficiently, or if the load increases while the vehicle slows down, the charger's voltage output may drop very near the battery's output. At these times, both the battery and the charger supply current to satisfy the vehicle load.

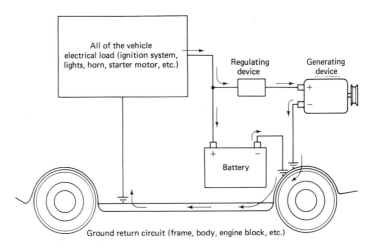

FIGURE 15–3. At increased loads and/or when the engine slows down, both the battery and the charger sometimes supply current (*Courtesy of Prentice-Hall, Inc.*).

BASIC PRINCIPLES

As indicated in previous chapters, whenever there is relative movement between a conductor and magnetic lines of force, the electrical balance of the conductor's atoms is upset. As the lines of force cut through the conductor's atomic structure, electron movement or current flow results.

The lines of force can be supplied by a permanent magnet or an electromagnet. However, since it is difficult to adjust the output of a permanent magnet (which is necessary to control voltage and current output), all automotive chargers use electromagnets.

In old-style generators there are two sets of windings: an outside stationary set called the field and an inside rotating set called the armature. The field windings make up the electromagnet. When current passes through the field windings, magnetic lines of force circle around and through the field shoes and frame.

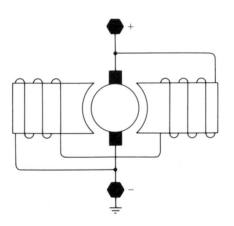

FIGURE 15–4.   Schematic of an old-style generator (*Courtesy of Prentice-Hall, Inc.*).

The interior, armature windings are mounted on a shaft connected to the generator drive pulley. When rotated by this pulley, the armature windings cut across the lines of force circling the field. As a result, electrons are put into motion in the armature windings. These electrons travel from the armature through the commutator segments to the brushes. Then the current goes to the battery and rest of the car's electrical system, eventually returning to the generator to form a complete circuit.

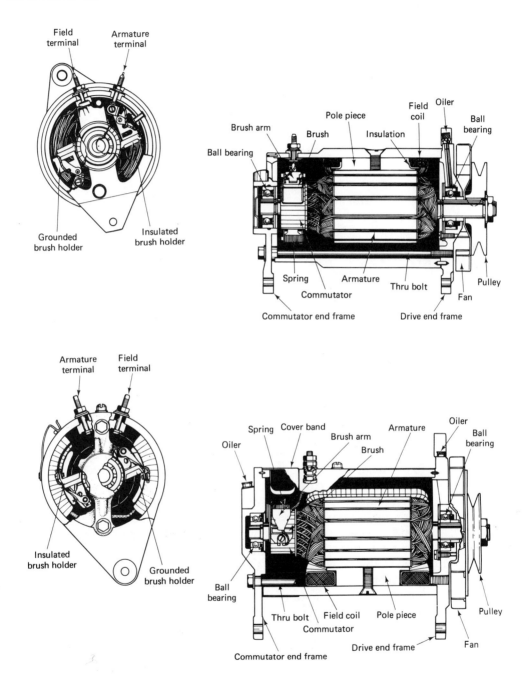

FIGURE 15-5. Components of old-style generators (*Courtesy of Prentice-Hall, Inc.*).

The output from the armature windings is controlled by varying the magnetic strength of the field windings.

The basic operating principles of an alternator and a generator are the same; that is, relative motion between lines of flux and a conductor induces voltage, which causes current flow in the conductor. The main differences are the parts that rotate and the parts that remain stationary. In an alternator, the conductor is stationary and the magnetic lines of force rotate. The effect is the same as in the generator—induced current flow due to relative motion between lines of force and conductor.

In an alternator, the outside conductor is called the stator, because it remains stationary. The center coil where the lines of force are created is called the rotor for obvious reasons.

CONSTRUCTION

The basic components of most alternators are the stator, rotor, two end frames to support these components, a shaft, and a variety of diodes and transistors.

Stator

The stator contains three main sets of windings wrapped in slots around a laminated, circular iron frame. A typical winding is made up of seven coils, which in turn are made up of a number of individual loops connected in series. The voltage produced in one loop is added to the next, and the voltage produced in one coil is added to the next.

Each main group of windings has two leads or ends, one lead where current enters the winding and one where current leaves. The leads are joined in two basic ways. In the so-called "Y" connection, one lead from each group of windings is joined in one common junction. The other leads branch out in a "Y" pattern from that single connection. In the "delta" or triangular connection, the lead at one end of a group of windings is joined to the lead at the other end of the next group, which is joined to the lead at the end of the next group, and so on. The "Y" configuration is the most common.

Each main group of windings occupies one third of the stator, or considering the stator as a circle with 360 degrees, each group takes up 120 degrees of the circle. This means the output from the stator is divided into three steps, or phases. As the rotor turns, first one phase will produce current, then the next, then the next. Alternators of this type are usually referred to as three-phase units.

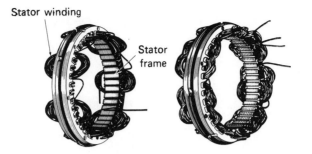

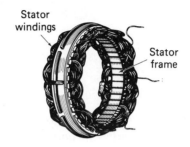

FIGURE 15–6. Alternator stator components (*Courtesy of Prentice-Hall, Inc.*).

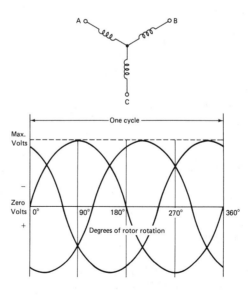

FIGURE 15–7. "Y" connection stator winding and diagram of voltage output (*Courtesy of Prentice-Hall, Inc.*).

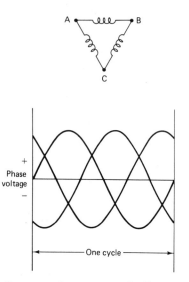

FIGURE 15-8. "Delta" connection stator winding and diagram of voltage output (*Courtesy of Prentice-Hall, Inc.*).

Rotors

A typical rotor has four main components. At the center is a drum-shaped coil of wire. Surrounding the coil are a pair of iron cups. The rims of the cups are formed into finger projections which interlace around the coil, like the joined fingers of two hands. A shaft is passed through the center of both cups and the coil. The shaft is mounted on bearings at both ends. Current for the rotor is delivered by brushes riding against insulated copper slip rings mounted on the rotor shaft. The slip rings are connected to the two ends of the rotor coil.

When current passes through the rotor coil, a magnetic field is set up in the windings. The lines of force snake in and out of the interlaced iron fingers surrounding the coil. As a result, the fingers become magnetized, one finger acquiring a north polarity, the next a south, and so on.

Diodes

Alternators, unlike generators, cannot use a split ring commutator to change AC to DC because the stator doesn't rotate. So, diodes are employed to change or rectify the current from AC to DC. Acting as one-way valves, the diodes switch the current flow back and forth so it only flows out of the alternator in one direction.

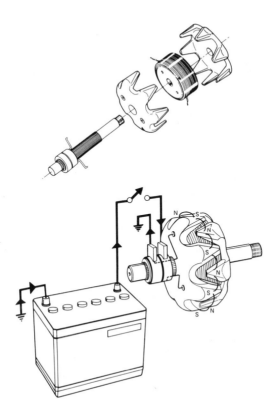

FIGURE 15-9. Alternator rotor (*Courtesy of Prentice-Hall, Inc.*).

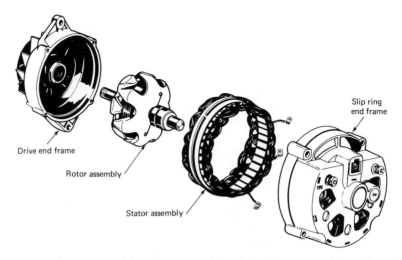

Drive end frame

Rotor assembly

Stator assembly

Slip ring
end frame

FIGURE 15-10. Complete alternator assembly (*Courtesy of Prentice-Hall, Inc.*).

An alternator usually has six diodes, three positive and three negative. The positive diodes are mounted in a "heat sink" on the back of the alternator. Heat from reverse bias voltage is conducted from the diodes to the metal heat sink and from there it is radiated to the surrounding air. The three negative diodes are attached directly to the end plate of the alternator. This provides a ground return circuit for the alternator output through these diodes.

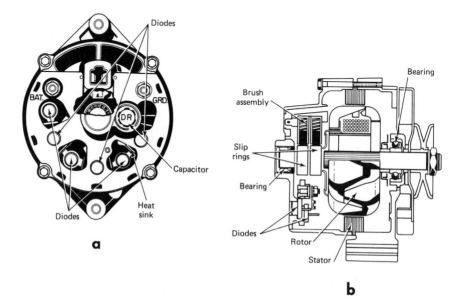

FIGURE 15–11.   **Note location of diodes in (A) (*Courtesy of Prentice-Hall, Inc.*).**

CURRENT FLOW (AC)

As the rotor poles turn within the stator, voltage is induced in the stator phases. During a single stator phase, current flows in one direction for 180 degrees of rotor rotation, then in the other direction for the remaining 180 degrees of rotation as opposite polarity poles pass by. Assuming a complete circuit exists, the voltage from any given phase reaches a peak at 90 and 270 degrees of rotation. At these points the magnetic region cutting across the stator windings is strongest. The voltage drops to zero at zero, 180 and 360 degrees of rotor rotation, when the magnetism is weakest.

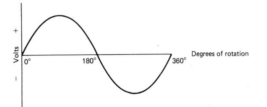

FIGURE 15–12. Unrectified, AC current flow from single stator phase (*Courtesy of Prentice-Hall, Inc.*).

CURRENT RECTIFICATION (AC TO DC)

Halfwave Rectification

Figure 15–13 is a schematic representation of one phase connected to one diode. The phase is shown as a single strand coiled into several loops. The diode symbol is a solid arrowhead pointing to a short vertical line. The direction the diode will allow current to flow is noted by the adjacent (+) and (−) signs. The diode allows current to flow when it is forward biased; that is, when the current enters positive to (+) or negative to (−). Otherwise it is reversed biased and will not allow current to flow. The output from the phase windings is also noted by (+) and (−) signs. These signs switch positions when opposite poles of the rotor pass by the phase.

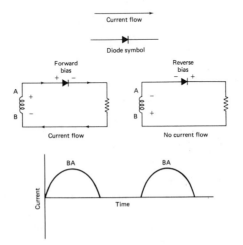

FIGURE 15–13. Half wave, rectified DC output from single stator phase (*Courtesy of Prentice-Hall, Inc.*).

For 180 degrees of rotor rotation, current will flow in a forward bias direction. It will pass through the diode and the attached circuit. For the other 180 degrees of rotor rotation, the current will be reversed biased and will not pass through the diode or the attached circuit.

The current flow diagram in Figure 15–13 shows that the single diode simply blocks out half the output from the phase. The entire lower wave is eliminated. Consequently, this is known as halfwave rectification.

Fullwave Rectification

Halfwave rectification is not desirable because too much of the alternator's output is eliminated. To solve this problem, engineers devised a system using the six diodes mentioned before. It is called fullwave rectification.

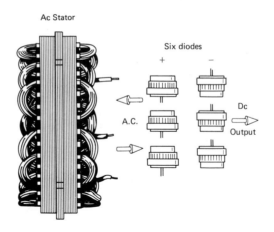

FIGURE 15–14.   These six diodes provide full wave rectification for the stator output (*Courtesy of Prentice-Hall, Inc.*).

Figures 15–15A through F illustrate the fullwave rectifying system. But before examining the flow paths, the parts will be noted. The diodes are known collectively as the rectifier bridge. The three alternator phases are identified "A," "B," and "C." Because they are connected in a "Y" pattern, one end of each phase is joined at the common center junction. The other ends are connected into the rectifier bridge.

As the rotor turns, current is induced in successive phases. In each phase, current is induced, first in one direction, then in the next direction.

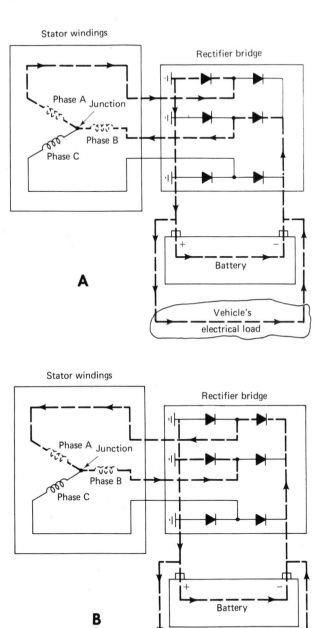

FIGURE 15-15. *(Courtesy of Prentice-Hall, Inc.).*

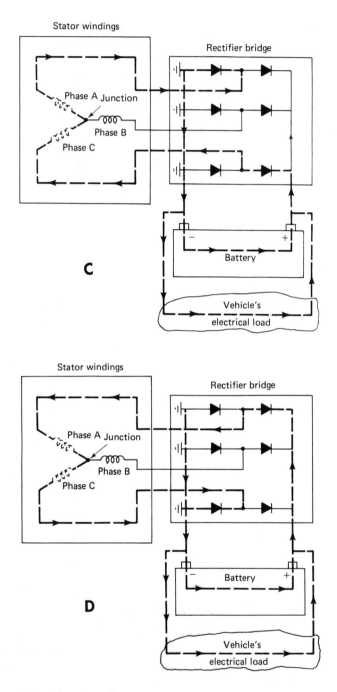

Figure 15-15 *(Continued)*

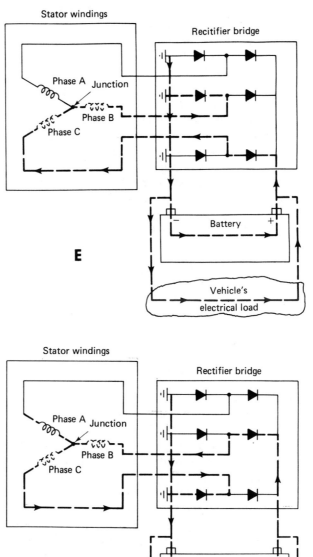

E

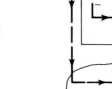

F

Figure 15-15   (Continued)

Figure 15-15 A shows the flow path in one direction from Phase A. Electrons are pushed (1) out of the loop, (2) through a forward bias diode, (3) to ground, (4) through the ground return circuit, (5) then to the negative pole of the battery. When alternator voltage (EMF) is greater than battery voltage (CEMF), electrons are pushed through the battery in a reverse direction for charging.

The electrons then go to the battery terminal on the rectifier bridge. From there, the electrons pass through a forward bias diode, go into phase winding "B," and return to phase winding "A" by way of the common connector "Y" junction. This completes the circuit.

The reverse output from Phase A is shown in Figure 15-15 B. In this case, electrons are pushed from phase A through the "Y" junction into the windings of phase B. From there, the electrons go on to the rectifying bridge, to ground, through the battery in a reverse direction for charging, then back to the rectifying bridge to the end of phase A to complete the circuit.

The flow paths for the other two phases are shown in Figures 15-15 C through F. Notice in each case that electrons are pushed from the phase windings, to the rectifying bridge, to ground, through the battery, then back to the rectifying bridge, and the phase windings to complete the circuit.

VOLTAGE REGULATION

The output from a generator typically requires three kinds of regulation. The cut-out relay prevents the battery from discharging through the generator; the current regulator prevents the generator from producing more current than the armature can handle, and the voltage regulator causes the generator to produce no more voltage than is needed to charge the battery and satisfy the vehicle load.

Alternators do not require all these regulating devices. A cut-out relay is not needed because battery current cannot pass through the reverse bias diodes in the rectifier bridge to reach the alternator.

A current regulator is not required because the alternator is not self-excited. Its operating current comes from the battery. This means the alternator cannot produce enough current to damage itself. Battery CEMF and the load in various operating circuits determine the current levels. At low battery CEMF, or load, the alternator current "draw" is high, and at higher battery CEMF or load, the current draw drops.

So, if current levels are determined by a push-pull relationship with the battery and other devices in the operating circuits, and if the diodes eliminate the need for a cut-out relay, the only control left is a voltage regulator.

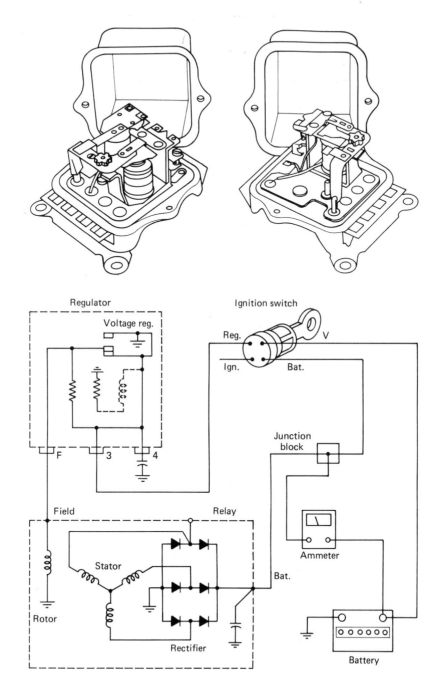

FIGURE 15–16. (*Courtesy of Prentice-Hall, Inc.*).

Modern automotive and light truck diesels employ a fully transistorized voltage regulating system. The regulator, which is usually incorporated into the alternator itself, is tested by techniques specified in the manufacturer's shop manual. If test results are negative, no repairs are attempted in most cases. The entire regulating unit is replaced.

CHAPTER 16

Diesel Cranking Systems

INTRODUCTION

Diesel starter motors and starting systems are essentially the same as those used in gasoline powered vehicles. As in so many cases, the only differences are due to the heavier starting loads imposed by diesel operation. This chapter briefly reviews the operation and components of automotive starting systems.

BASIC SYSTEM CIRCUITRY

The starter motor is part of the starting system, which is part of the car's complete electrical system. Figure 16–1 shows some of the main elements in a simplified cranking system circuit: the battery, the start switch, a neutral safety switch, a cranking motor control switch, the cranking motor itself, and the necessary cables and wires to connect these components.

There are usually two main paths along which current may flow in a typical cranking system circuit. One path, represented by the heavy black lines, follows the thick cable running from the battery through the cranking motor control switch to the cranking motor. The other path follows the smaller wire from the start switch to the neutral safety switch and on to the cranking motor control switch.

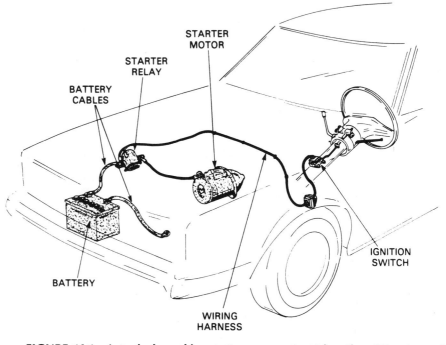

FIGURE 16.1 A typical cranking system component location (*Courtesy of Ford Motor Company of Canada, Ltd.*).

The thick cable provides a low resistance path for the heavy current flow required to operate the starter motor. The start switch and neutral safety switch *could* be included in this main operating circuit, but it might be dangerous. So, the start and neutral safety switches are connected by high resistance wires to reduce current flow and pressure.

The parallel branch in which the start and neutral safety switches are included is basically a signal or control circuit for the starter motor control switch. When the start and neutral safety switches are closed, current flows to the motor control. The current energizes an electromagnet that closes the starter motor control switch that allows current to flow along the heavy cable to the starter motor. Current continues to flow until the two signal switches are opened and the signal circuit is broken. Then the electromagnet breaks the main operating circuit to the starter motor.

STARTING SYSTEM BASIC MECHANICAL COMPONENTS

The basic mechanical components of a simplified starter system are shown schematically in Figure 16-2. There is a motor for supplying torque or turning force, a pinion or small gear attached to the motor's drive shaft, a flywheel ring gear, and a drive engagement mechanism of some kind. The motor turns the pinion, which engages the flywheel ring gear, which turns the engine over. The drive engagement mechanism moves the pinion into mesh with the flywheel when the engine is being cranked and moves it out of mesh after the engine starts running.

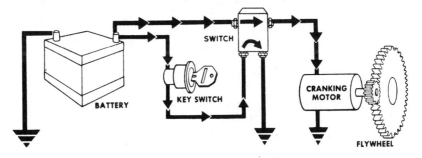

FIGURE 16-2. **Cranking system schematic and gear reduction (*Courtesy of General Motors Corporation*).**

BASIC ELECTRIC MOTOR CONSTRUCTION

A typical starter motor contains field pole shoes and windings, the starter housing (or field frame), the armature assembly, the brushes, and plain bearing bushings supporting either end of the armature (see Figure 16-3).

The pole shoes are constructed of soft iron for increased magnetic permeability. They are attached to the inside of the field frame by bevel head screws.

The armature is also constructed from soft iron laminated with other materials for strength. Special slots are machined into the armature for the armature loops. The loops are made of heavy copper strips for reduced resistance.

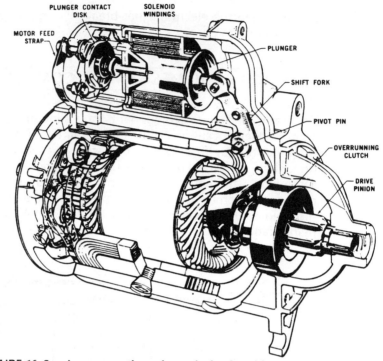

FIGURE 16-3.   A cross-section of a typical solenoid type of cranking motor (*Courtesy of Ford Motor Company of Canada, Ltd.*).

The ends of the armature strips are attached to the commutator segments. The commutator is constructed of wedge-shaped copper pieces pressed into heat-resistant mica insulating material. The insulation separates the copper segments from one another and from the armature drive shaft, which passes through the center of the commutator. The commutator is machined to nearly perfect roundness on a lathe so the brushes will follow true as the armature rotates.

The brushes are usually constructed of copper-carbon compounds. Most starter motors have four brushes. Two are grounded to the frame or end plate, and two are insulated from the frame.

Starter motor armatures are usually supported at both ends by bronze-brass bushings. However, some heavy duty motors have an extra, center bushing for added support.

Most starter motors are series wound. In a typical example, current flows first (1) to the field windings, (2) then to the insulated brushes, (3) then through the commutator segment and the (4) armature winding contacting the brushes at that instant, (5) then out through the grounded brushes.

PERFORMANCE CHARACTERISTICS

Starter motors use electromotive force (EMF) to operate. The torque and speed of the motor depend on the amount of current and the strength of the magnetic fields. The strength of the magnetic fields, in turn, is determined by the number of windings, the nature of the core material, plus the amount of current flowing. The amount of current flowing depends on the supply of EMF from the battery and the resistance encountered in the motor circuits.

In series wound motors, current flow and the resultant torque are greater at lower motor speeds. This means that as the motor first starts to turn over, when the load is greatest, the current and torque will be high. Then as the motor speeds up, the current flow and torque will drop.

One of the reasons current falls off as the motor speeds up is the CEMF (counter electromotive force) generated in the armature windings. As the armature rotates within the field's magnetic lines of force, an EMF is generated in the armature counter to the EMF from the battery. The motor, in effect, acts as a generator, inducing current flow as the armature windings cut across the field's flux lines. The motor reaches its maximum speed when the EMF from the battery is balanced by the resistance in the windings plus the CEMF from the armature.

The electric starter motor is a high performance device. It produces considerable horsepower and torque under heavy loading. To reduce current flow and possible overheating, the motor is designed to operate at high speeds where current flow is least. For this reason, the gear ratio between the pinion and the flywheel ring gear is 15:1 to 20:1. In other words, the starter motor turns 15 to 20 times as fast as the engine. However, even with these precautions, the starter motor will overheat and suffer damage if allowed to operate much more than 30 seconds at one time.

MOTOR CIRCUITS

Most starter motors in use today have similar wiring circuits. Figure 16-4 shows a typical example. It has four pole shoes in the field assembly, two of which contain coils. The shoes wrapped with coils are magnetic north, and unwrapped shoes, magnetic south. The field flux lines go out from the north poles, pass through the armature, go into the south poles, then return to the north poles through the field frame assembly. (Remember: magnetic lines of force must follow a continuous flow path. Also, remember that flux lines PREFER to pass through mate-

rials of high magnetic permeability—in this case, the iron shoes, armature and field frame.)

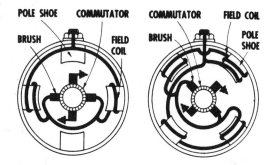

FIGURE 16–4.   Schematic wiring diagram of two-field, four-pole, four-
brush, series-wound cranking motor (at left). Schematic
wiring diagram of four-brush, four-field, four-pole series-
wound cranking motor (at right). (*Courtesy of General
Motors Corporation*).

The 4-pole, 2-coil motor just described acts in much the same way
as a 4-pole, 4-coil motor, except that it offers less resistance to electrical
flow since there are fewer windings. This is one of the most common
motor circuits. However, others are in use. There are 4-coil, 4-pole
motors; motors with 6 coils wrapped around 6 poles; motors with a series wound shunt coil to limit top speed, etc.   Figure 16–5 shows one
of these other motor configurations.

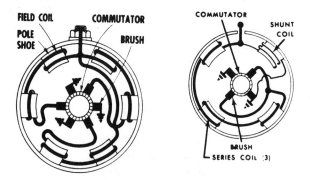

FIGURE 16–5.   Schematic wiring diagram of a six-field, six-pole, six-brush,
series-wound cranking motor (at left). Shunt coil controls
excessive armature speed at light load, allowing heavier field
coil winding for more torque (at right). (*Courtesy of General
Motors Corporation*).

REASONS FOR
CONTROL CIRCUITS AND
DRIVE ENGAGEMENT MECHANISM

If it were not for two fundamental facts, a cranking system need not be much more than a motor, start switch, pinion, and flywheel ring gear.

First, the heavy current required to operate the cranking motor (as much as 100 amps) should not be routed directly through the start switch. This might involve some danger, plus the cables in the dash would have to be excessively large.

Second, the pinion cannot be allowed to remain engaged with the flywheel after the engine cranks.

Given a gear ratio between the pinion and flywheel of 15 or 20 to 1, if they remained engaged after the engine was running at 1,000 RPM, the starter motor would be driven by the engine at speeds of 15,000 to 20,000 RPM. The centrifugal forces at such speeds would quickly throw the windings out of the armature and destroy the motor.

It is because of these basic facts that the starting system requires some special features—a control circuit and associated switches and a drive engagement mechanism to move the pinion in and out of mesh with the flywheel ring gear. The following sections describe these special features of the starting system.

STARTER DRIVES

There are two basic kinds of devices used to move the pinion in and out of mesh. One is the Bendix, or inertia drive (and its several variations). The other is the overrunning clutch drive.

Bendix Type Inertia Drive

In standard Bendix drives, the pinion is fabricated around the outside of a hollow barrel. The inside of the barrel has coarse cut, screw threads. These threads are matched by similar threads on a sleeve assembly. The barrel is loosely threaded onto the sleeve, and the sleeve is mounted onto the end of the armature drive shaft.

Torque from the armature is transmitted via a shock absorbing spring and drive head to the sleeve. This causes the sleeve to rotate. However, at first, the barrel and pinion do not rotate with the sleeve. The loosely fitting barrel has a weight on one side to increase its inertia (the natural tendency of any object to resist change in velocity). So, the sleeve rotates inside the barrel. As a result, the barrel screws itself down the length of the sleeve to the end where it engages the flywheel gear. At that point, the barrel locks in place and begins to transmit torque from the armature. The so-called "Bendix spring" connecting the sleeve to the end of the armature shaft absorbs the engagement shock with the flywheel.

As soon as the engine starts running on its own, it will rotate faster than the armature. This causes the pinion to be screwed back down the sleeve and out of mesh with the flywheel. Should the pinion be driven back down with excessive force, a small overriding clutch is activated to prevent damage to the starter motor.

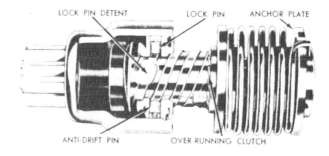

FIGURE 16-6.   Bendix type of starter drive.

Bendix Barrel Drive.   The barrel drive is similar to the standard Bendix drive. The principal differences are these: the pinion is usually mounted on the end of the barrel; it is generally smaller (for higher gear ratios); and it works directly off screw threads at the end of the armature shaft instead of through an intervening sleeve assembly.

Bendix Folo-Thru Drives.   This drive is also similar to the Bendix barrel drive. The principal differences are the addition of a "detent" pin and a "detent" clutch.

The detent pin locks the barrel in place on the screwshaft. This locking action takes place after the pinion barrel has moved to the end of the screw shaft, when the shaft is turning rapidly. Centrifugal force throws the pin into engagement between the barrel and the screwshaft.

The detent clutch connects the two sections of the screwshaft. If the engine should run faster than the armature shaft, the clutch disengages, thus protecting the motor from damage.

Inboard or Outboard Drives. In many Bendix drives the pinion is screwed "outward" from the motor to engage the flywheel. These are called "outboard" drives. In other cases the pinion is driven inward toward the motor. This is called an inboard drive.

Overrunning Clutch Drive

The overrunning clutch drive operates on an entirely different principle from Bendix drives. Instead of using an inertia operated screw to move the pinion in and out of mesh with the flywheel ring gear, the pinion is moved by a shift linkage of some kind. In fact, the overrunning clutch itself might not be considered the most important element in this kind of drive. The control element that operates the pinion shift linkage could be considered more important since it actually moves the pinion into mesh with the flywheel.

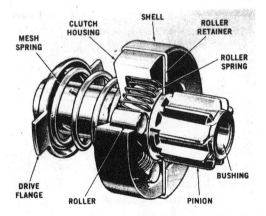

FIGURE 16–7.   A cross-section of an overrunning clutch type of starter drive (*Courtesy of Ford Motor Company of Canada, Ltd.*).

The overrunning clutch, like the clutch in the folo-thru drive, is primarily a safety device separating the armature drive shaft from the pinion. A sleeve and hollow shell assembly make up one-half of the clutch. They are splined onto the armature drive shaft. The pinion and a collar comprise the other half of the clutch. The pinion is mounted onto the collar, which fits inside the shell.

Positive engagement between the collar and the shell is provided by spring-loaded rollers located in slots within the shell. When the armature is turning slower than the engine, the rollers jam into contact with the sleeve causing the connected pinion to transmit torque to the engine. But when the engine runs faster than the armature, the rollers are driven up their ramps to the wider side of the slots. Then the rollers act as bearings so the collar can spin freely inside the shell and in so doing not transmit any damaging torque back to the motor.

Some overrunning clutch drives use a system of oval-shaped rollers called "sprags." These sprags rock back and forth into engagement with the pinion collar.

The shift linkage used to move the pinion in and out of engagement with the flywheel ring gear may be manually operated or controlled by an electric solenoid. Most starters today use the solenoid control. It will be described presently.

Dyer Drives

These drives are often used in heavy duty truck or industrial applications. They combine features of both Bendix type and overrunning clutch drives. Like the overrunning clutch drive, the pinion is meshed with the flywheel by a shift lever. And like the inertia drive, the pinion is spun out of mesh along screw threads once the engine operates faster than the starter motor.

CONTROL SWITCHES

The switches that control the flow of high amperage current to the starter motor should not be considered apart from the starter drive. Bendix drives almost always employ a magnetic switch. And overrunning clutch drives generally use a solenoid switch to control current flow as well as move the pinion in and out of mesh with the flywheel ring gear.

MAGNETIC SWITCHES

Used in Bendix drives, these switches consist primarily of a coil, a plunger with a disc connected at one end, and electrical contacts attached to the main operating circuit. The switch may be located in any convenient place along the main operating circuit cable.

When the operator turns the start switch to a closed position, the signal circuit sends current to the magnetic switch windings. This creates a magnetic field around the windings. The magnetic field pulls the plunger inside the coil. When the plunger moves, the contact disc on the end joins the contacts on the two sides of the main operating circuit. High amperage current then flows through disc and contacts to the starter motor.

So long as the operator holds the switch key in the crank position, the circuit will remain closed, the magnetic switch windings will remain energized, and the main operating circuit will allow current to flow to the starter motor. However, when the operator releases the switch, the signal circuit will be broken and the magnetic field will collapse. Then, the spring-loaded plunger will move the contact disc away from the main circuit contacts, breaking that circuit and turning the starter motor off.

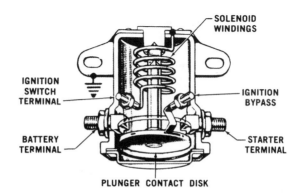

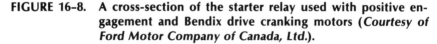

FIGURE 16–8.   A cross-section of the starter relay used with positive engagement and Bendix drive cranking motors (*Courtesy of Ford Motor Company of Canada, Ltd.*).

SOLENOID SWITCH

The plunger in the solenoid switch has two jobs. It must close the main operating circuit so current will flow to the starter motor. It must also operate the shift linkage that moves the pinion in and out of mesh. (Remember, the solenoid is used primarily with overrunning clutch drives.)

To do these two jobs, two sets of windings are required. One, the pull-in winding, has the heavy duty job of shifting the pinion into mesh. The other, the hold-in winding, holds the plunger disc against the main circuit contacts after the pinion has engaged. The pull-in winding is fabricated from heavy wire to reduce resistance and produces the strongest field. The hold-in winding has more turns of finer wire.

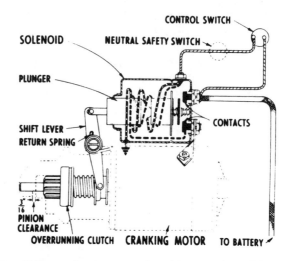

FIGURE 16-9.   Shift mechanism on solenoid type of cranking motor (*Courtesy of General Motors Corporation*).

CHAPTER 17

Solenoids, Switches, and Relays

INTRODUCTION

More and more mechanical operations are being monitored or controlled by electrical devices. Part of the reason is that electrical equipment is sometimes less expensive than mechanical hardware. A more compelling reason is the speed, precision, and versatility possible with electrical controls. Modern automotive and small truck diesels are no exception to this trend. The following paragraphs briefly review the prime agents of electro-mechanical control: solenoids, switches, and relays.

SOLENOIDS

A solenoid translates electrical impulses into mechanical action. It is considered a remote control device because the solenoid can be located one place and the solenoid switch somewhere else; the former where the mechanical action is needed, and the latter where the control decisions are made (either manually or automatically).

A typical solenoid consists of a coil (or coils) of wire wrapped around a hollow tube. When electrical current flows through the coil, a strong magnetic field is generated. The field, as it flows around and through the hollow tube, causes the solenoid to act like a magnet with the ability to attract iron-based objects. However, unlike permanent magnets, a solenoid can draw objects inside its hollow core. These objects, often referred to as core rods, are attached by a linkage system to the mechanical device operated by the solenoid.

Current flow to a solenoid coil can be determined automatically by various means: temperature or pressure sensitive switches, limit switches responding to the movement of some other devices, etc. Current flow can also be controlled by manually operated switches. The starter solenoid shown in Figure 17–1 is controlled by the manually operated key switch. Current flow to the GM diesel throttle control solenoid, pictured in Figure 17–2, is determined by a temperature sensor.

SWITCHES

Various kinds of electrical switches are used: simple units that control the flow to a single device or complex switches used to route current flow to a network of circuits. Some switches are manually operated; others are automatically opened or closed.

Diesels contain most of the switches found in gasoline powered vehicles and then some. Reviewing briefly, they include the following:

1. ***Headlight Switches*** direct current flow to the headlights, parking lights, taillights, etc. Besides providing the usual push-pull notches, many headlight switches also include a twist-operated rheostat to adjust the intensity of the interior and instrument lights.

2. ***Dimmer Switches*** also regulate current flow to the headlights.

3. ***Signal Switches*** control current flow to the turn signals, brake lights, instrument panel information/warning lights (including the glow plug signal lights), fuel gauge, etc. Except for the turn signal, these switches are automatically operated by devices responding to pressure, temperature, the position of some other element, etc.

4. ***Safety Switches*** route current flow to the door, key, and brake buzzers and in some cases to the starter to prevent cranking if the automatic transmission selector lever is not in park or neutral. These switches are also automatically controlled.

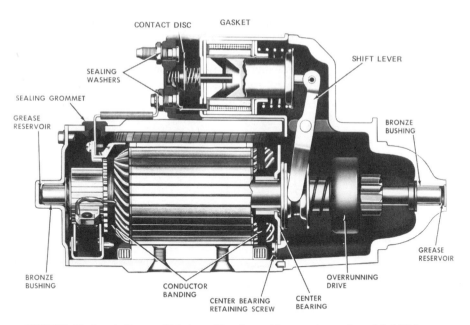

FIGURE 17-1. A General Motors diesel cranking motor, solenoid shift lever, and contact disc (*Courtesy of Oldsmobile Division, General Motors Corporation*).

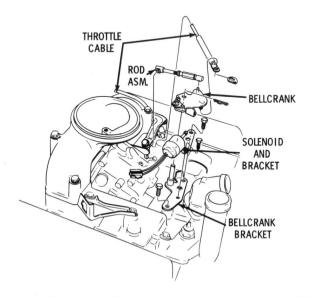

FIGURE 17-2. A diesel throttle solenoid (*Courtesy of Oldsmobile Division, General Motors Corporation*).

5. **Key Switches** control current flow to most other switches and circuits in the vehicle. In gasoline vehicles, the key switch controls current flow to the ignition system. Since diesels don't have ignition systems, the key switch acts as a central control for the glow plug circuit.

Figure 17–3 shows several types of mechanically and manually operated switches.

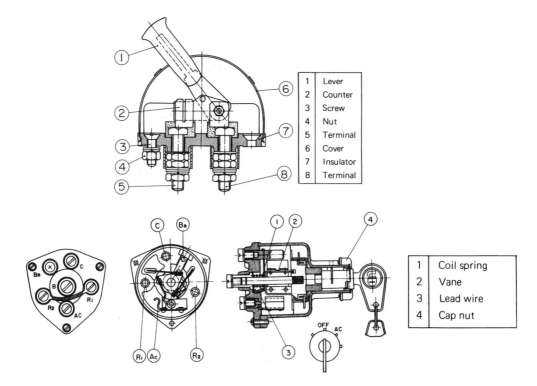

FIGURE 17–3. "Main master switch" and starter switch (*Courtesy of Nissan Diesel Motors Ltd./Marubeni America Corporation*).

RELAYS

Like solenoids, relays are designed to translate electrical impulses into mechanical action. However, instead of a moveable iron core, relays have a fixed core. And instead of using mechanical action to operate a linkage of some kind, relays are used as remote control switches. In a typical example, an operator engages a small manual switch that allows current to flow to a relay. Once energized, the relay then closes (or opens) a much larger switch, which controls a much greater current flow.

Such an arrangement offers several advantages. It allows the operator's switch to be smaller; it eliminates the need to run heavy cables into the operator's area, and it reduces the danger of shocks to the operator from current flowing through the heavier cables.

The smaller switches that control current flow to relays can either be manually operated or activated by various kinds of sensors. Some of the devices using relays in diesel powered cars and small trucks include the:

1. Horn
2. Lights
3. Charge indicator
4. Glow plugs
5. Air conditioner clutch
6. Heater motor

Figure 17–4 shows the glow plug relays used on a GM diesel. A typical relay schematic is pictured in Figure 17–5.

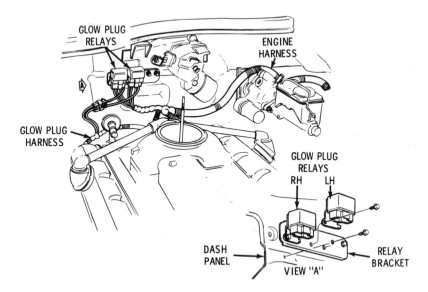

FIGURE 17–4.   A glow plug relay (*Courtesy of Oldsmobile Division, General Motors Corporation*).

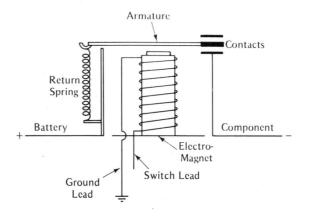

FIGURE 17-5.   A schematic of a typical relay design.

CHAPTER 18

Starting Aids

INTRODUCTION/BACKGROUND

Diesel engines, owing to the nature of the fuels used and to the self-ignition combustion process, do not start as easily as gasoline engines. This is especially true at temperatures below 55 or 60 °F.

Large trucks, heavy equipment, and stationary diesels use several methods to get around this problem. Some employ coolant heaters as either standard or optional equipment. These heaters contain an electric heating element incorporated into the block at a convenient place for warming the coolant liquid. Operating current is usually obtained from a standard 110V outlet. This method is practical for vehicles that, once started, remain in more or less continuous operation. The engine does not cool down and therefore does not need to be hooked back up to an electrical outlet for reheating. Obviously, the technique is not practical for private vehicles. A 110V outlet would be needed everywhere diesel cars and small trucks are parked.

Another approach for cold starting large industrial diesels is to spray highly combustible fuels, like ether, into the air induction system. The most sophisticated examples of this technique inject a measured amount of starting fuel through an atomizer located in the induction system. Another common (and more dangerous) practice is to simply spray ether into the air intake from an aerosol can.

Given the inconvenience of coolant heaters and the inconvenience as well as hazards of starting fluids, manufacturers of automotive and small truck diesels (as well as many large truck diesels) rely on another cold start method, the glow plug.

GLOW PLUGS

A glow plug is simply a small, electrically operated heater. The tip, which projects into the combustion chamber, is the actual heating element. It not only warms the cylinder walls and combustion area, it (most importantly) heats the incoming fuel. This helps vaporization and makes combustion possible in cold conditions.

Glow plugs are threaded into the cylinder head much like spark plugs. Generally, each cylinder contains one glow plug. Figure 18-1 pictures a typical glow plug installation.

Power for the glow plugs comes from the vehicle's own electrical system. After passing through a control circuit of some kind, the current flows in a parallel path through a heavy cable or bus bar to the individual glow plugs. Figure 18-2 is a schematic of a simplified glow plug circuit. An actual GM glow plug circuit is shown in Figure 18-3. Figure 18-4 shows a Nissan circuit.

Given the job glow plugs must perform, the current draw is necessarily very high. This is one reason why the batteries used in diesels are usually more powerful than their gasoline engine counterparts. As noted in a previous chapter, it is also why some manufacturers use two batteries wired in parallel.

However, even with the additional battery power, other steps must be taken to minimize the electrical drain resulting from glow plug operation. Some manufacturers reduce total current flow by "pulsing" the glow plugs. In other words, instead of maintaining continuous current flow during the starting period, the flow is periodically interrupted. It is also common to limit glow plug operation with a timer or thermostatic control. The glow plugs are shut off after a specified period of time or after a particular temperature has been reached.

Glow plugs can be checked with a test light or an ohmmeter. Since glow plugs are heater units with low resistance, test lights are especially useful for quickly finding out if a glow plug's internal wiring is complete.

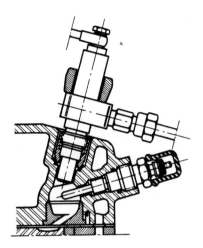

FIGURE 18-1. A swirl chamber with the injector and glow plug shown (*Courtesy of Peugeot*).

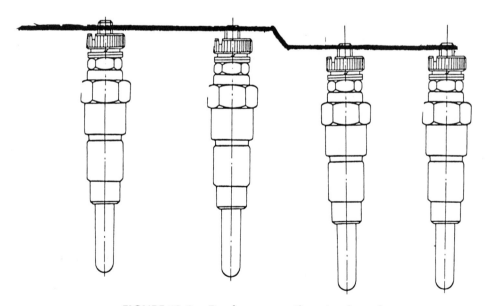

FIGURE 18-2. Bus bar connections to glow plugs.

SECOND TYPE DIESEL GLOW PLUG ELECTRICAL CONTROL

1	2	3		5	6
				┌─ ENGINE RUNNING ─┐	
IGN. SWITCH -"OFF"-	IGN. SWITCH -"RUN"-	IGN. SWITCH -"RUN"-	IGN. SWITCH -"START"-	IGN. SWITCH -"RUN"-	IGN. SWITCH -"RUN"-
WAIT LAMP - OFF	WAIT LAMP - ON	WAIT LAMP - OFF	WAIT LAMP - OFF	WAIT LAMP - OFF	WAIT LAMP - OFF
START LAMP - OFF	START LAMP - OFF	START LAMP - ON	START LAMP - ON	START LAMP - OFF	START LAMP - OFF
GLOW PLUGS - OFF	GLOW PLUGS - ON	GLOW PLUGS - ON	GLOW PLUGS - ON	GLOW PLUGS - ON	GLOW PLUGS - OFF
		↓		↓	
		SEE NOTE 1		SEE NOTE 2	

NOTE 1: If the ignition is left in the "Run" position without starting the engine, the glow plugs will continue to pulse on/off until batteries run down. (About 4 hours when coolant switch is open.)

NOTE 3: Do not manually energize or by-pass the glow plug relay as glow plugs will be damaged instantly.

NOTE 2: Glow plugs will pulse on/off for about 30 seconds after engine starts. Then turn off and remain off as long as engine temperature is above about 120° F (49° C).

NOTE 4: Diodes prevent glow plug operation when the engine is warm (above 120°) the engine is not running and key is in RUN.

IMPORTANT: Do Not use more than a 2—3 candle power test light when making circuit checks.

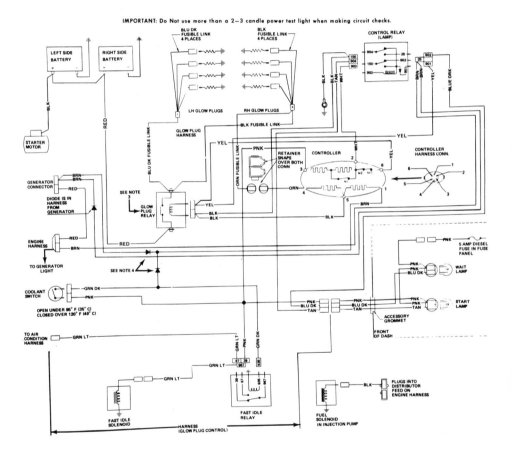

FIGURE 18-3.   A glow plug electrical control circuit (*Courtesy of Oldsmobile Division, General Motors Corporation*).

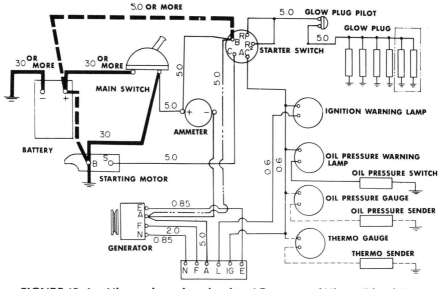

FIGURE 18-4. Nissan glow plug circuitry (*Courtesy of Nissan Diesel Motors Ltd./Marubeni America Corporation*).

CHAPTER 19

Checking Cylinder Compression

INTRODUCTION

As the piston rises during the compression stroke, the pressure of the gases trapped in the combustion chamber increases proportionally. This is true for both gasoline and diesel powered engines. However, the pressure increase is greater in diesel engines because of the higher compression ratios necessary to ignite diesel fuel.

The actual pounds per square inch exerted by the air when the piston is at TDC varies from one diesel manufacturer to another. The principal advantage of knowing the pressure, both specified and actual, is being able to diagnose certain engine conditions. Compression figures that fall below factory specified levels usually indicate a problem with the valve train, the piston/ring/cylinder system, or the head gasket.

COMPRESSION TESTER

Various factory-made and aftermarket compression gauges (as pictured in Figure 19-1) are in use. Most have these features in common: (1) a dial face calibrated in PSI or kg/cm^2, (2) a pressure sensitive element of some kind usually located in the dial face housing, (3) a tube connected to the housing at one end and a fitting at the other end, (4) fitting adapters allowing the tester to be used on various kinds of engines, and (5) a valve to control the flow of high pressure air to the sensing element.

FIGURE 19-1.   A "Denis Godard" compression test gauge (*Courtesy of Peugeot*).

GENERAL TEST PROCEDURES

As noted at the end of this chapter, the exact procedures for checking diesel engine compression vary somewhat from manufacturer to manufacturer. However, certain steps are common for all procedures.

1. Before checking compression, make sure the battery is fully charged. A weak battery might not turn the engine over fast enough to give an accurate compression reading.

2. Some manufacturers specify that the glow plugs be removed to check the compression; others specify removal of the injector nozzles. Before removing either, clean away surrounding dirt so it won't fall into the cylinders.

3. Screw the tester fitting into the glow plug or injector nozzle opening. (Or, if required, first install an adapter, then screw the tester fitting into the adapter.)

4. Put the gearshift or selector level in park or neutral. Set the emergency brake.

5. Using either a remote starter or with the aid of a helper sitting behind the wheel, engage the starter motor. As the engine starts to turn over, open the tester valve.

6. Let the engine turn over for four or five seconds. Some mechanics listen for six audible "puffs" of air coming from the openings into the other cylinders.

7. While the engine is turning over, watch the reading on the tester dial. It should build up to a certain level and stay there. Make a note of the reading.

8. Check the other cylinders in the same way.

ANALYZING COMPRESSION RESULTS

Compression results are compared against the factory specifications. Low readings indicate problems in one of three areas.

1. *Valve Train.* Many valve problems, including damaged valves and valve seats, broken valve springs, and burned valve faces or seats will prevent the valves from seating properly. This in turn will allow air to escape from the combustion chamber and will reduce compression readings.

2. *Piston/Ring/Cylinder System.* Worn or damaged pistons, rings, or cylinder walls will result in air leaks and a subsequent loss of compression.

3. *Blown Head Gasket.* This too will reduce compression.

Specialized test equipment will pinpoint the specific trouble area.

SOME SPECIFIC DIESEL TEST PROCEDURES

Following are outlines of the compression test procedures recommended by several manufacturers of diesel engines. Consult the actual factory shop manual if you are performing these tests yourself.

GM V-8 Diesel

1. Remove the air cleaner and install the special air crossover cover.

2. Disconnect the wire from the fuel solenoid terminal of the injection pump.

3. Disconnect the wires from the glow plugs; remove all the glow plugs.
4. Screw a compression gauge into the glow plug hole of the cylinder being checked.
5. Crank the engine over.
6. Note the compression reading.
7. **Results:** The lowest cylinder reading should not be less than 70% of the highest and no cylinder reading should be less than 275 PSI.

Volkswagen Rabbit Diesel

(See Figure 19-2.)

1. Remove the wire from the stop control on the injection pump. Insulate the wire end.
2. Remove the injector pipes.
3. Remove the injector nozzles.
4. Screw in the special VW 1323/2 adapter and the VW 1323 compression tester. Place the old heat shield between the adapter and the head.
5. Crank the engine and check the compression.
6. After checking the compression, make sure to install new heat shields between the cylinder head and the injectors.
7. **Results:** New engines should read 34 kg/cm^2 (483 PSI). Older engines should read no less than 28 kg/cm^2 (398 PSI). The difference between individual cylinders should be no more than 5 kg/cm^2 (71 PSI).

Nissan Small Truck Diesel

(See Figure 19-3.)

1. Loosen the nozzle and injection pump cap nuts and separate the nozzle line assembly.
2. Remove all nozzle holder assemblies.
3. Screw the compression gauge into the cylinder head, making sure to use the proper adapter. The tightening torque for the adapter is 55.7 to 57.9 lb. ft.
4. Disengage the clutch for safety.
5. Set the control lever of the injection pump at zero delivery.

FIGURE 19-2. Testing compression on a Volkswagen diesel (*Courtesy of Volkswagen*).

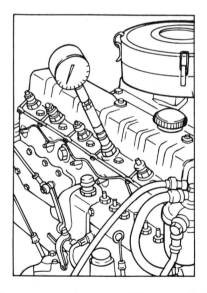

FIGURE 19-3. Testing compression on a Nissan small truck diesel engine (*Courtesy of Nissan Diesel Motors Ltd./Marubeni America Corporation*).

6. Close the main disconnect switch; turn the starter key and crank the engine. Normally the maximum reading is achieved in five seconds.

7. *Results:* Readings below 19 kg/cm² (270 PSI) at crank speeds below 200 RPM indicate insufficient compression. A difference between maximum and minimum readings of more than 4 kg/cm² (57 PSI) is excessive.

Peugeot Diesel

1. Make sure the engine is at operating temperature (approximately 80°C or 178°F).

2. Remove the injectors.

3. Block the injection pump stop control in the cut-off position.

4. Install a "Denis Godard" pressure gauge in the first cylinder being checked.

5. Tighten the knurled screw.

6. Engage the starter for approximately four seconds. The engine should turn over at about 300 RPM.

7. Check the compression.

8. After the compression test has been completed, make sure to use new sockets when reconnecting the fuel lines.

9. *Results:* With the engine at normal operating conditions, the compression should measure 40 to 45 BARS with the Godard gauge (or approximately 362 PSI). If measured with a moto-meter gauge, the reading should be between 20 to 25 BARS under the same conditions.

CHAPTER 20

Checking Injection Pump Timing

INTRODUCTION

Timing the exact moment of diesel fuel injection serves the same purpose as timing the spark in a gasoline engine. It determines when the combustion process starts. In gasoline engines, timing is usually checked with a strobe light hooked up to shine every time a single spark plug fires. The light "freezes" the motion of a rotating reference mark on the end of the crankshaft. The position of the mark is compared to a nearby stationary mark. Timing adjustments show up as shifts in the position of the rotating mark.

Diesels are not normally timed in this manner. Instead of "freezing" a rotating reference mark with respect to a stationary mark, a combination of stationary marks (or readings on a dial indicator) are compared while the engine is not running. Only basic timing is checked, not timing advance. The following paragraphs briefly outline the procedures recommended by several diesel manufacturers for setting basic timing. Before doing any actual work, consult a factory shop manual.

GENERAL MOTORS

The GM rotary injection pump is properly timed when the marks on the top of the injection pump adapter and flange line up. If they don't, loosen the three pump retaining nuts and rotate the pump about its drive shaft axis until the marks are aligned (see Figure 20–1).

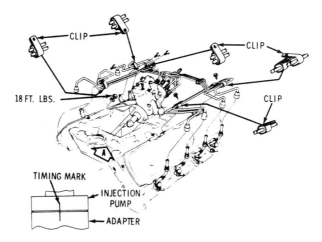

FIGURE 20–1.   General Motors timing marks (*Courtesy of General Motors Corporation*).

NISSAN

Basic timing of the Nissan, Bosch-style, multi-plunger pump can be quickly checked by comparing the reference marks on the injection pump and front end plate (see Figure 20–2). The timing is satisfactory if the marks line up. If they don't, loosen the pump hold-down bolts and shift the pump until the marks are aligned.

To verify that the timing marks themselves are accurate (usually done when the pump is reinstalled after service), perform this procedure.

1. Install the pump and timer assembly on the engine. Make sure the gear backlash is correct. Connect the fuel supply line to the pump, but don't connect the high pressure nozzle lines yet. Remove the delivery spring from the number one pump outlet and reinstall the valve holder at the correct torque (see Figure 20–3).

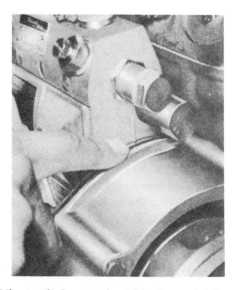

FIGURE 20-2. Nissan timing marks (*Courtesy of Nissan Diesel Motors Ltd./Marubeni America Corporation*).

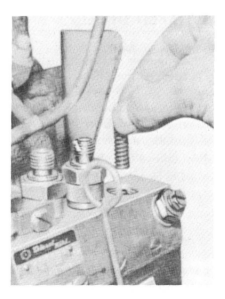

FIGURE 20-3. Removing delivery spring prior to verifying timing (*Courtesy of Nissan Diesel Motors Ltd./Marubeni America Corporation*).

2. Make sure the injector pump hold-down bolts are just loose enough so that you can rock the pump back and forth.

3. Rotate the engine in a normal direction until the first mark on the crank pulley lines up with the reference mark on the gear case. This corresponds to 20 degrees BTDC of the number one piston (see Figure 20-4).

4. Use the priming pump to supply the injection pump with fuel.

5. Rock the pump all the way toward the engine.

6. While watching fuel coming from the number one outlet, slowly pull the pump back away from the engine. Stop at the instant the fuel ceases to flow. This is the beginning of injection for the number one cylinder.

7. Lock the injection pump in place.

8. Look at the reference marks on the injection pump and front end plate. They should be aligned. If they are not, make a new mark on the front end plate. These marks can be used henceforth for routine timing checks.

9. Assemble the delivery spring removed previously and complete the rest of pump assembly procedure (see Figure 20-5).

VOLKSWAGEN RABBIT DIESEL

When the engine is installed, the basic timing of the Volkswagen rotary pump is checked by comparing two sets of reference marks.

1. Set the number one piston at TDC by rotating the engine until the TDC mark on the flywheel lines up with the raised boss on the bell housing.

2. Remove the plug from the pump cover.

3. Check to see if the marks on the pump and mounting plate line up (see Figure 20-6).

If the engine has been removed from the vehicle, follow this procedure:

1. Place a setting bar against the flywheel and turn the engine until TDC is indicated (see Figure 20-7).

2. Remove the plug from the pump cover.

3. Using the proper adapter, insert a dial gauge in the pump cover hole (see Figure 20-8).

FIGURE 20-4.   Indicating the Number One cylinder at 20° BTDC (*Courtesy of Nissan Diesel Motors Ltd./Marubeni America Corporation*).

FIGURE 20-5.   Installing nozzle lines (*Courtesy of Nissan Diesel Motors Ltd./Marubeni America Corporation*).

FIGURE 20-6.   Volkswagen timing marks (with pump cover plug removed) (*Courtesy of Volkswagen*).

FIGURE 20-7.   Setting TDC with the setting bar (*Courtesy of Volkswagen*).

FIGURE 20-8.   Dial indicator installed in pump cover hole (to set timing with engine removed) (*Courtesy of Volkswagen*).

4.   Adjust the dial indicator until it reads a preload value of 2.5 mm.

5.   Turn the engine in the opposite direction of normal rotation until the indicator needle stops moving.

6.   Preload the gauge to 1 mm, then set the needle to zero.

7.   Turn the engine in the direction of normal rotation until it once again reads TDC on the setting bar.

8.   The gauge should be at .83 mm. If it isn't, loosen the bolts on the mounting plate and support and make the proper adjustments.

CHAPTER 21

Checking Injector Nozzles

INTRODUCTION

The shop manuals of some diesel manufacturers provide procedures for testing injector nozzles. These procedures usually test nozzle opening pressures and spray patterns. In some cases, advice is given for checking the noise made by the injectors, since certain sounds are associated with correct operation.

TEST EQUIPMENT

Although various types of test devices are available and are used in slightly different ways depending on the manufacturer (Figures 21-1 and 21-2), most have these features in common:

1. A hand-operated pump mounted on the test bench. This pump bears a certain resemblance to a hydraulic jack.
2. A pressure gauge located on the pump.

3. An enclosure to protect the operator from the nozzle spray. In some cases, this enclosure is made of clear plexiglass.
4. A high pressure line leading from the pump to the nozzle being tested.
5. A nozzle holder located in the enclosure.
6. Shut-off valve(s) to protect the meter when the pump pressure is suddenly released.

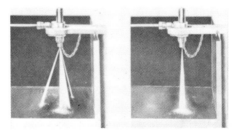

FIGURE 21-1. Peugeot injector nozzle test apparatus (*Courtesy of Peugeot*).

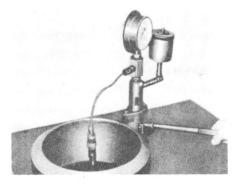

FIGURE 21-2. Nissan injector nozzle test apparatus (*Courtesy of Nissan Diesel Motors Ltd./Marubeni America Corporation*).

CHECKING INJECTOR OPENING PRESSURE

Follow these general steps to determine injector opening pressure.

1. Remove the injectors from the engine.
2. Make sure both the injectors and the tester are clean.
3. Install the first injector to be checked on the tester injector holder.
4. Fill the pump supply chamber with special viscosity test oil (kerosene). Use clean, fresh oil.
5. Operate the pump arm at a rate of about one stroke per second. **Note/Caution:** Avoid contact with the high pressure spray coming from the injector tip. Wear eye protection.
6. Observe the reading on the tester dial. The pointer will oscillate slightly, so take an average of the high and low readings.
7. Compare the reading to the manufacturer's specifications. (In one example, this is 100 kg/cm^2 or 1,422.3 PSI.)
8. The opening pressure of some injectors is adjusted by adding or removing nozzle spring shims. If the nozzle being checked has these provisions and the opening pressure is not correct, make the necessary shim adjustments.
9. Before removing an injector from the tester, turn off the valve to the pressure dial. Then, after the injector has been disconnected, slowly open the valve to let the needle gradually return to its stop. Rapid pressure drops can damage the meter (see Figure 21–3).
10. Check the other injectors, making adjustments as necessary (and where possible) to ensure that the opening pressure is the same for each.

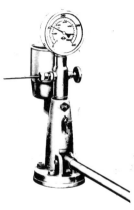

FIGURE 21-3.   A test pump (*Courtesy of Peugeot*).

CHECKING INJECTOR LEAKAGE

Pump the tester until it reads about 20 kg/cm^2 less than the previously observed opening pressure. Let the pressure stay at this level for ten seconds. No fuel should drip from the nozzle tip during the test period.

CHECKING SPRAY PATTERN
AND INJECTOR "HUM"

Follow these general procedures

1. Shut off the valve to the tester pressure gauge.
2. Pump the tester at about one stroke per second. The fuel should come from the injector in a thin, even jet.
3. Gradually increase the pumping rate. At about 1 to 2 strokes per second, a properly working injector should produce a characteristic "humming" sound.
4. Continue to increase the pumping rate to about 4 to 6 strokes per second. The "humming" noise should diminish and the fuel should exit from the nozzle in an even, cone-shaped mist (see Figure 21-4).

Note/Caution. If different kinds of injectors from various engines are being checked, make sure the correct injectors go back to their proper engines. Peugeot claims that using injectors and/or holders with the wrong pump can lead to excessive fuel consumption, smoky exhaust, rapid deterioration of the injectors and possibly the engine itself.

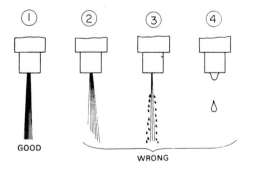

FIGURE 21-4. Spray patterns (*Courtesy of Nissan Diesel Motors Ltd./ Marubeni America Corporation*).

CHAPTER 22

Checking the Fuel Pump

INTRODUCTION

As noted in Chapter 9, diesel engines contain at least two fuel pumping devices, the fuel injection pump and the delivery or supply pump. The fuel injection pump is responsible for providing measured quantities of high pressure fuel to the injector nozzles. The delivery pump supplies fuel at much lower pressures to the injector pump (see Figure 22-1).

Checking the fuel injector pump is a specialized task usually performed by service centers with the tools and trained personnel to handle the job. However, some manufacturers do provide procedures for checking the delivery pump. The following paragraphs briefly review some of these tests.

GENERAL MOTORS

At the time of this writing, GM uses the same supply pump on both gasoline and diesel powered V-8's. Consequently, this pump, which is mounted on the block and operated by an eccentric on the camshaft, can be tested in the same way for both applications. Perform these general steps:

Test Supply Pressure and Quantity

1. Remove the outlet hose from the pump.
2. Attach the tester, using adapters if necessary.
3. Pinch or clip the outlet hose of the tester so that no fuel can escape.
4. Start the engine and let it run at approximately 1200 RPM.
5. Note the pressure reading.
6. Open the outlet hose of the tester, allowing the fuel to run into a graduated cylinder for ten seconds.
7. ***Results:*** A pump in good working order should deliver approximately one pint of fuel in ten seconds when the engine is running at 1200 RPM. The pressure should be approximately 4½ to 6 PSI.

Test Supply Pump Vacuum

To check the vacuum pulled by the supply pump, first disconnect the fuel inlet line to the pump. Then connect a vacuum gauge to the inlet port. The reading should be about 12 inches of mercury.

INTEGRAL PUMPS

Many Bosch-type, multi-plunger pumps combine the injection and supply pumps in one unit. Where the proper connections can be made, these pumps are tested in a manner similar to the previously described procedure. The proper result will depend on the manufacturer's specification. For instance, one Nissan small truck diesel is designed to deliver 300 cc or more of fuel every 15 seconds when the engine is running at 1000 RPM. The pressure in this example should be 1.6 kg/cm^2 (22.78 PSI).

Some simple, shop-level tests are also available for checking these pumps when they have been removed from the engine. Here are two tests for checking (1) overall performance and (2) airtightness (see Figure 22–2).

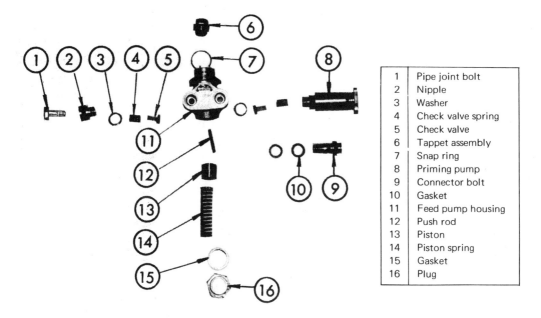

1	Pipe joint bolt
2	Nipple
3	Washer
4	Check valve spring
5	Check valve
6	Tappet assembly
7	Snap ring
8	Priming pump
9	Connector bolt
10	Gasket
11	Feed pump housing
12	Push rod
13	Piston
14	Piston spring
15	Gasket
16	Plug

FIGURE 22-1.   Exploded view of a fuel feed pump (*Courtesy of Nissan Diesel Motors Ltd./Marubeni America Corporation*).

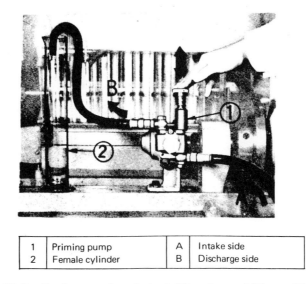

1	Priming pump	A	Intake side
2	Female cylinder	B	Discharge side

FIGURE 22-2.   Fuel pump bench test (*Courtesy of Nissan Diesel Motors Ltd./Marubeni America Corporation*).

Test Overall Performance

1. Clean the supply pump and mount it on a test bench.
2. Connect a pipe to the pump's intake port. Let the end of the pipe hang down into a container of clean diesel fuel located approximately one meter (39.37") below the level of the pump.
3. Connect another pipe to the pump's outlet port. Place the end of this pipe into a container.
4. Work the hand operated primer valve at a rate of about 80 strokes per minute.
5. **Results:** If the pump starts delivering fuel from the outlet within one minute, it is probably O.K.

Check Airtightness

1. Plug the pump's outlet port.
2. While applying about 25 to 28 PSI of pressure to inlet port, immerse the pump into a container of light oil (kerosene).
3. **Results:** The pump is sufficiently airtight if no bubbles or only occasional bubbles appear. However, a continuous discharge of bubbles from any location indicates an airleak. If this happens, determine the source of the bubbles and attempt to fix the leak.

CHAPTER 23

Checking Exhaust Smoke

INTRODUCTION

The color and quantity of smoke coming out of a diesel exhaust pipe directly relates to what takes place inside the engine. This fact has long been recognized by mechanics who work on heavy equipment diesels. One of the first things these mechanics do when troubleshooting a problem is to check the exhaust smoke. It can help them trace faults back to the fuel or air induction systems, the rings, etc. This chapter reviews some of the basic relationships between diesel exhaust smoke and engine conditions.

NORMAL DIESEL SMOKE

Most diesels will occasionally produce visible exhaust smoke. A tank of low grade fuel may cause smoking. Some black, sooty smoke may appear under heavy load or acceleration or when the engine is first started. White smoke may also appear when cranking a cold engine in cold weather. However, if the engine smokes much more than this, or if the smoking is accompanied by poor performance, then a problem is indicated.

THREE TYPES OF SMOKE

Diesel exhaust smoke appears in three basic colors. Each relates to certain engine conditions.

Black, Sooty Smoke

This kind of smoke contains unburned or partially burned fuel. It can be produced by an otherwise sound engine if something happens to momentarily disrupt the air/fuel balance or the combustion process; e.g., if the engine speed or load suddenly changes, if the fuel isn't the proper quality, or if the engine isn't hot enough to completely burn the fuel. However, if the black, sooty smoke is produced continuously, one or more of the following conditions is likely to be present.

Low Compression. If the compression is too low, air in the combustion chamber will not be squeezed enough and will not become hot enough to ignite the fuel properly. Low compression can be caused by stuck or broken rings, valves that fail to close properly, or worn pistons and cylinders.

Fuel System Malfunctions. Black smoke will also be produced if too much fuel goes into the combustion chamber or if the fuel enters at the wrong time. Some problems that can contribute to improper fuel delivery include malfunctioning delivery nozzles, poorly timed fuel injection pumps, etc. The procedure for timing fuel injection is given in Chapter 20. Injectors can be checked by loosening the injector fuel lines, one at a time, while the engine is running. The technique is similar to selectively removing spark plug wires in a gasoline powered engine. Disconnecting a properly working injector (or spark plug) should cause engine performance to fall off. If it doesn't, the injector wasn't working well to begin with. If the smoking condition improves when the injector is removed, it is also safe to assume that the injector was part of that problem. *Note/Caution:* Be especially careful when removing injector lines. Direct the fuel spray away from any possible sources of ignition. Wear eye protection. Keep a fire extinguisher handy.

Air Induction Problems. If enough air isn't present, fuel cannot burn properly and black smoke will be produced. This can be caused by a clogged air cleaner or by an obstruction in any of the air flow passages.

Blue Smoke

As in gasoline engines, blue smoke indicates the presence of lubricating oils in the combustion chamber. This in turn suggests cylinder and/or ring wear or some other condition causing excessive lubrication. Blue smoke should never come from a properly working engine.

White Smoke

White smoke usually means water has entered the combustion chamber. As noted before, a certain amount of white smoke can be expected when a cold engine is cranked in cold weather (or sometimes in warm, damp weather). That is because the atmospheric moisture drawn inside the engine isn't completely vaporized. The water droplets remain large enough long enough to condense into a visible cloud when they encounter the cool air at the end of the tail pipe. How long the white smoke appears depends on the water content of the air and the temperature of the air and engine. In extremely cold weather, a visible vapor cloud may be evident for many miles after the vehicle is cranked. It is like the condensation trails produced by jet planes flying in the cold upper reaches of the atmosphere.

However, if the engine constantly produces white smoke and if the coolant needs frequent refilling, there is probably a water leak into the combustion chamber. In most cases, this will be due to a head gasket failure. However, water jacket leaks cannot be completely ruled out.

NO-START SMOKE

It is not necessary for an engine to run to produce telltale smoke. Sometimes an engine that fails to start will give off clouds of black, blue, or white smoke during the cranking effort. If that happens, the engine defect is likely to be the same kind that produces smoke when the engine is running, only more severe.

SEVERAL KINDS OF SMOKE AT ONE TIME

If an engine has a smoking problem, it is likely that more than one defect is present. For instance, if an engine has stuck rings, it might produce both blue smoke from burning lubricants and black smoke from unburned fuel. The result would be blue/grey smoke. Various combinations and shades of smoke are possible, each signifying a slightly different set of conditions. Association with experienced mechanics is the best way to learn to judge subtle differences between smoke.

CHAPTER 24

Engine Cylinder Block Service

INTRODUCTION

As noted in an earlier chapter, the block is the structural heart of the engine. It contains the explosive power of the combustion process as well as the twisting forces required to convert combustion energy into useful mechanical power. Because a modern block is so strong and so well designed, it hardly ever wears so badly or fails so completely that total replacement is necessary. However, with sufficient age or mileage, service to the block becomes likely. And, as in most other aspects of major service, work done to the block is often accompanied by repairs to other major components (see Figures 24-1 and 24-2).

PROBLEMS

Gross Physical Damage

In the past, it wasn't uncommon to hear the phrase, "throw a rod." It meant that a piston rod had broken or worked loose at one of the bearing points. A broken rod was often accompanied by a hole in the side of the block where the rod had gone through. Although much less likely now (particularly in the case of relatively low RPM diesel engines), it is still possible. It could conceivably take place as the result of age and wear, a poorly done rebuilding job, or extreme and unusual stress.

Also fairly common in the past were cracked blocks or popped freeze plugs due to frozen engine coolants. Although still possible, this kind of damage is unlikely with modern coolants and cooling systems.

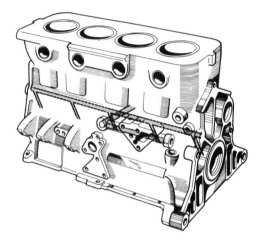

FIGURE 24-1. A cylinder block (*Courtesy of Peugeot*).

More common now are cracks or warping resulting from mechanical or heat stresses. For instance, operating an old engine at high speeds could cause cracks in areas that had suffered metal fatigue. Stress cracks could also appear in the block of a newly rebuilt engine whose components were improperly aligned or torqued together.

The block can warp if the engine becomes so hot that it expands (and then shrinks) unevenly. Warpage can also result from improperly torqued studs (particularly head studs) anchored in the block (see Figure 24-3).

Cylinder Wall Damage/Wear

One of the most common problems with the block is wear or damage to the cylinder walls. Dirt or grit is often the culprit, resulting in shallow gouges or scratches. Not only does this allow lubricants to escape it

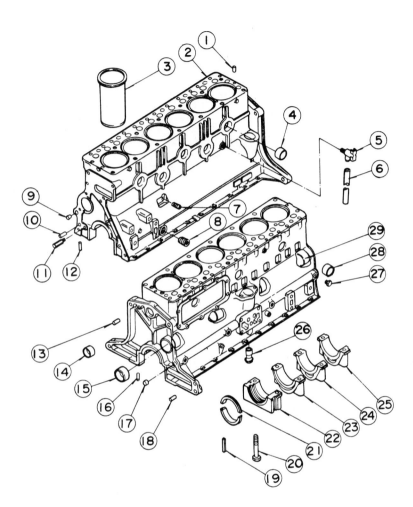

1	Dowel pin	11	Oil jet	21	Oil seal
2	Cylinder block	12	Plug	22	Main bearing cap (No. 4)
3	Cylinder liner	13	Straight pin	23	Main bearing cap (No. 3)
4	Plug	14	Plug	24	Main bearing cap (No. 2)
5	Water drain cock	15	Plug	25	Main bearing cap (No. 1)
6	Vinyl hose	16	Straight pin	26	Bushing
7	Plug	17	Plug	27	Plug
8	Plug	18	Straight pin	28	Plug
9	Plug	19	Oil seal	29	Camshaft bushing
10	Dowel pin	20	Main bearing cap bolt		

FIGURE 24–2. Cylinder block assembly (*Courtesy of Nissan Diesel Motors Ltd./Marubeni America Corporation*).

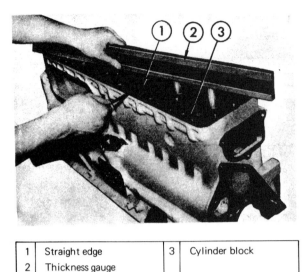

1	Straight edge	3	Cylinder block
2	Thickness gauge		

FIGURE 24-3. **Using a straightedge to check the top of the block for warping (*Courtesy of Nissan Diesel Motors Ltd./Marubeni America Corporation*).**

also provides a passage for combustion gases to "blow by" the piston rings and contaminate the oil supply.

In addition to wear of this nature, broken rings can damage the cylinder walls. So can a piston pin lock that has worked itself loose.

Clogged Cooling Passages

Chemical deposits or rust scales can accumulate sufficiently to block off the flow of coolant liquid through the water jacket. This can then result in overheating, which can cause the block to become warped.

BLOCK SERVICE

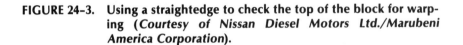

NOTE/CAUTION

The cylinder block is the heaviest part of the engine and extreme care must be taken when it is removed. Consult the manufacturer's recommendations.

Honing Cylinder Walls

Some piston ring manufacturers recommend that the natural glaze that builds up on cylinder walls be removed before new rings are installed. The glaze is removed by the honing process. Honing is also required if the cylinder walls are scuffed or wavy (see Figure 24–4).

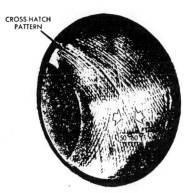

CROSS-HATCH
PATTERN

FIGURE 24–4.   The cross-hatch pattern of a honed cylinder (*Courtesy of Gould, Inc.*).

A cylinder hone is a special grit block attached to the chuck of a conventional electric drill. The drill is held at a slight angle as it is moved up and down the cylinder so that a cross-hatch pattern appears on the surface of the cylinder wall. If the honing is done with the crankshaft in place, care must be taken to keep debris out of the crankcase. After honing is complete, the cylinder walls should be thoroughly cleaned with soap and water (see Figures 24–5 and 24–6).

Boring Cylinder Walls

If the cylinder walls are deeply scratched, reboring may be required (see Figures 24–7 and 24–8). The first step in this process is to attach a device called a boring bar to the deck or top part of the block. After being centered over the cylinder, the boring bar cutter is passed up and down in the cylinder until the desired diameter is achieved. After that, the cylinder walls are honed.

The exact diameter of the new cylinder bore is determined by the depth of the damage to the cylinder walls and the size of the new piston to be installed. Standard, oversize pistons are used whenever a cylinder is rebored.

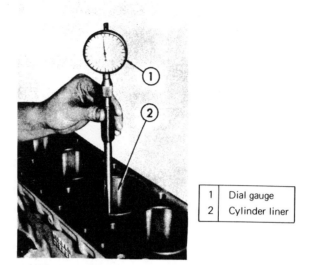

| 1 | Dial gauge |
| 2 | Cylinder liner |

FIGURE 24-5.   Measuring a cylinder bore (*Courtesy of Nissan Diesel Motors Ltd./Marubeni America Corporation*).

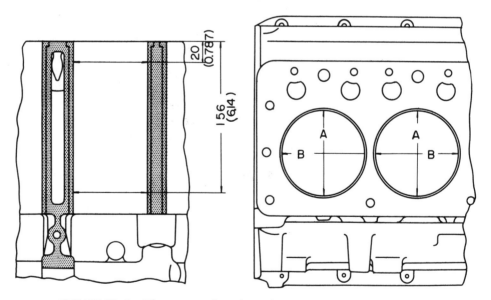

FIGURE 24-6.   The proper location of measurements (*Courtesy of Nissan Diesel Motors Ltd./Marubeni America Corporation*).

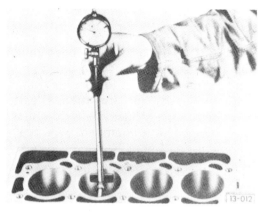

FIGURE 24-7. Checking clearance and wear (*Courtesy of Volkswagen*).

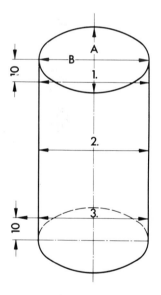

FIGURE 24-8. Proper measuring points (*Courtesy of Volkswagen*).

Removing Cylinder Ridges

After a period of use, ridges may build up around the top of the cylinders. These ridges must be removed using special ridge removal tools before the pistons can be lifted from the top of the cylinder bores.

Replacing Cylinder Sleeves

If a cylinder wall has actually been punctured, or if it has been too deeply gouged, it is sometimes possible to insert a new sleeve in the cylinder. The first step is to bore the cylinder to the correct diameter to accept the sleeve. Then, after the sleeve is pressed into place, it is bored and honed to the proper size for the new piston. Several diesel automotive engines are equipped with replaceable sleeves as standard equipment (see Figure 24–9).

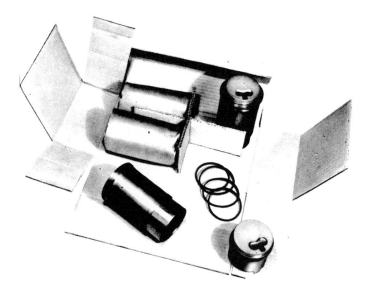

FIGURE 24–9.   Cylinder liners or sleeves (*Courtesy of Peugeot*).

Milling the Cylinder Deck

It is sometimes possible to correct a warped block by milling or cutting the top of the cylinder deck until it is once again perpendicular to the cylinders. If warpage is even suspected, the cylinders should never be bored until after the deck has been milled. However, the manufacturer's recommendations should be checked before any of this work is done. Some specify that a warped block be replaced rather than repaired.

Milling is done on a device similar to a large woodworking jointer. The block deck is passed back and forth over the cutting head until the deck intersects the cylinders at the proper angle.

Line Boring Main Bearings

If the main bearing bores are warped or misaligned, they can be restored to tolerance by the line boring process. A line borer is like a large, fixed drill press that passes a cutting element down the exact center line of the crankshaft. Line boring can be done in one of two ways: with special undersize main bearings already bolted in the bearing bores, or with just the bearing caps bolted into position.

DIESEL TIPS

The preceding suggestions apply equally to diesel and gasoline engines. However, because diesels are subjected to greater pressure, factory procedures and tolerances should be strictly observed. In particular:

1. Be especially careful to use the correct rings.
2. Closely inspect the cylinders for wear.
3. Make sure gasket surfaces are not scored or distorted.
4. Before reassembly, wash away all abrasives from the block with warm water and soap solutions.

CHAPTER 25

Crankshaft and Camshaft Service

INTRODUCTION

Internal combustion engines, including diesels, contain various "brains" for controlling the actions of other components. The crankshaft and camshaft are mechanical brains, the former determining when and how far each piston travels, the latter controlling valve opening and closing. The crankshaft also performs the vital role of converting the up and down movement of the pistons into force applied in a circular manner.

Because these two shafts are so important, and because repairs are so expensive and time-consuming, they are designed to last a long time. However, both can wear or become damaged. For instance, both shafts are very sensitive to lubrication failures. The crankshaft, especially in diesels, operates under a great deal of stress (see Figure 25–1).

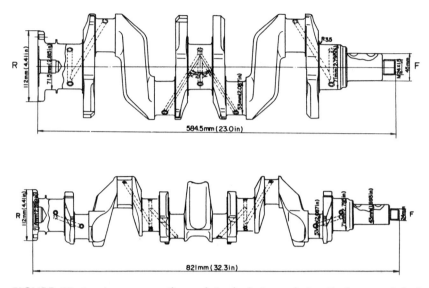

FIGURE 25-1. A cross-section of typical 4- and 6-cylinder crankshafts (*Courtesy of Nissan Diesel Motors Ltd./Marubeni America Corporation*).

CRANKSHAFT PROBLEMS

Journal Wear/Damage

The journals are the most likely candidates for crankshaft wear or damage. Not only are stresses concentrated at the journals but the journals are also very sensitive to any disruption in the flow of clean lubricating oil. Following are some typical journal related problems:

1. *Pitting.* Acids and harmful combustion by-products in the lubricating oils can eat holes in the journal surface.
2. *Nicks and Cuts.* This kind of damage can occur during a reassembly process if the mechanic accidentally bumps the crankshaft against some other part.
3. *Scoring and Scratching.* Such damage usually shows up as grooves cut around the circumference of the journal. It can happen if dirt or debris becomes trapped between the journal and the bearing surface.
4. *Wearing Out of Round.* As the piston and connecting rod assembly move down in the cylinder during the power stroke, pressure is concentrated on one side of the crankshaft journal. Pressure is concentrated on other sides of the journal during

the other strokes. After millions of revolutions, these forces can cause the journal to assume a slightly out-of-round or oval shape. The condition is checked by measuring the journal diameter at several points with a sensitive micrometer (see Figures 25-2A, B, and C).

5. *Wearing to a Taper.* Due to uneven application of forces, journals can also wear in a tapered manner; e.g., somewhat like a cone. This condition is also checked by measuring the journal diameter in several places with a micrometer.

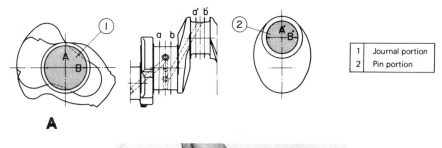

A

| 1 | Journal portion |
| 2 | Pin portion |

B

C

FIGURE 25-2. Measurement locations and procedures for measuring a crankshaft (*Courtesy of Nissan Diesel Motors Ltd./Marubeni America Corporation*).

Bent Crankshafts

Crankshafts are hardly ever bent permanently out of shape. However, all mechanical devices, including crankshafts, are designed to operate within certain tolerances. A crankshaft, for instance, is made of steel that performs within certain limits; it is balanced only so much and it is installed in a block bore that has been machined to certain maximum and minimum values.

A crankshaft fabricated in this manner will probably last for many millions of revolutions, flexing and bending a certain amount, but always returning to shape. However, with time, wear, and possibly some unusual abuse or extreme operating condition, the tolerances can be exceeded. It is at this point that the crankshaft, subjected to uneven stresses and operating in an uneven manner, may become bent.

A crankshaft suspected of being bent is checked with a dial micrometer (see Figures 25–3A, B). First, the crankshaft is placed in a cradle made of vee-shaped blocks [or it is left in the engine block with the center bearing cap(s) removed]. Then the dial micrometer is placed against each of the journals as the crankshaft is rotated in the cradle. Any irregularities will show up as fluctuations on the dial. If the run-out readings remain within specified limits, the crankshaft is properly aligned. If the fluctuations exceed the allowed limits, the crankshaft is bent (see Figure 25–4).

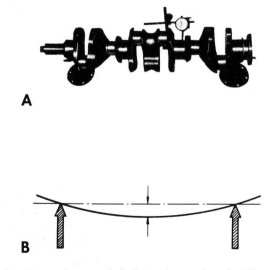

A

B

FIGURE 25–3. Measuring crankshaft to detect bends (*Courtesy of Nissan Diesel Motors Ltd./Marubeni America Corporation*).

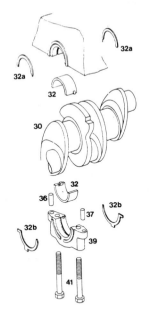

32   Bearing shell
32a Thrust washer, crankcase
32b Thrust washer, bearing cap
36   Dowel pin, 10 × 6 × 16 mm
37   Dowel pin, 8 × 6 × 16 mm
39   Bearing cap
41   Screw, M 12 × 75

FIGURE 25-4.   A cross-section of a portion of a Mercedes crankshaft (*Courtesy of Mercedes-Benz*).

Cracks

Some of the same problems that cause bent crankshafts can also result in cracks. Cracks severe enough to cause damage in a normally operated engine can usually be seen with the naked eye (after the crankshaft has been thoroughly cleaned).

CRANKSHAFT SERVICE

Crankshaft service is generally limited to either polishing the journals if the shaft is otherwise sound, or grinding the journals if the surfaces are pitted, cut, scored, worn out-of-round, or worn to a taper. Grinding is done on a special machine with the journal diameter reduced enough so that standard, replacement bushings can be used.

If the crankshaft is cracked or bent, or if one of the other conditions noted above is severe enough to require grinding in excess of specifications, the crankshaft is replaced.

CAMSHAFT PROBLEMS/SERVICE

Although camshafts can suffer the same damage as crankshafts, problems are usually limited to the cam lobes. The most common damage occurs at the nose where the valve train exerts the greatest force. However, unless a lubrication failure occurs, a camshaft will usually last the life of an engine. If the camshaft does wear beyond normal limits, it is usually less expensive to replace it than to attempt repairs (see Figures 25–5 and 25–6 A, B).

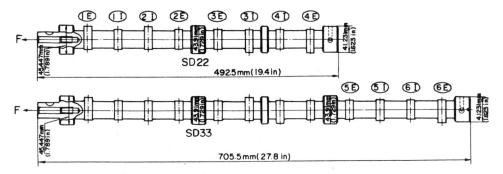

FIGURE 25–5. Typical camshaft journal dimensions (*Courtesy of Nissan Diesel Motors Ltd./Marubeni America Corporation*).

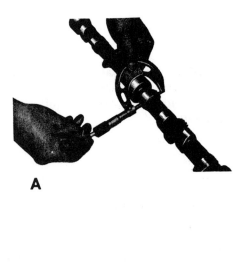

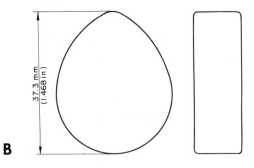

FIGURE 25–6. Measuring the camshaft (*Courtesy of Nissan Diesel Motors Ltd./Marubeni America Corporation*).

CHAPTER 26

Piston, Piston Ring, and Connecting Rod Service

INTRODUCTION

The pistons, along with the valves, cylinder head, and cylinder walls, are subjected to some of the most violent action in an internal combustion engine. Expanding gases push with explosive energy on the piston head. Side forces drag the skirt against the cylinder wall as the piston moves up and down. Consequently, the pistons and associated components are among the most common objects of major service. This is true for diesel as well as gasoline powered engines. Although diesel pistons are often more sturdy than their gasoline engine counterparts, they are also subjected to greater stresses and pressures (see Figure 26-1).

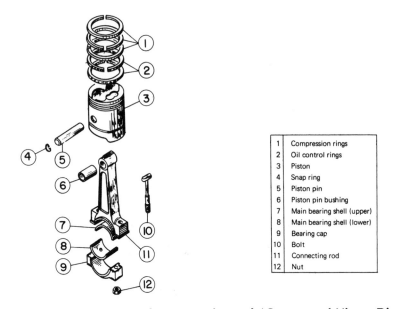

1	Compression rings
2	Oil control rings
3	Piston
4	Snap ring
5	Piston pin
6	Piston pin bushing
7	Main bearing shell (upper)
8	Main bearing shell (lower)
9	Bearing cap
10	Bolt
11	Connecting rod
12	Nut

FIGURE 26-1.   A piston and a connecting rod (*Courtesy of Nissan Diesel Motors Ltd./Marubeni America Corporation*).

PROBLEMS WITH PISTONS AND ASSOCIATED COMPONENTS

Ring Damage

The most common piston-related problem is a ring failure of some kind. Rings can become worn due to high mileage, improper lubrication, or improper break-in. Rings can break for the same reasons. Breakage can also occur when excessive wear in the piston ring grooves allows the rings to shift position and become subjected to uneven pressures.

Condensation/Corrosion Damage

Moisture sometimes gets into the combustion chamber, either as a result of a coolant leak, or more commonly, when the engine temperature remains low enough (long enough) for atmospheric moisture drawn into the engine to condense. In either case, the moisture can combine with combustion by-products to form a corrosive mixture. This corrosive material can cause pitting in aluminum pistons.

Combustion/Heat Damage

Uneven combustion conditions can damage pistons. Preignition is one example. In diesels it can occur if the fuel ignites too soon or if fuel left over from a previous cycle begins burning before the new charge enters the combustion chamber. Preignition results in extremely hot pockets of burning fuel. These hot spots can soften the top of a piston sufficiently to allow normal combustion chamber pressures to push a hole through the piston surface.

Another form of abnormal combustion is detonation or knocking. Collectively these terms describe the fuel explosions that occur near the edge of the combustion chamber, away from the main flame front proceeding out from the injector nozzle. The audible detonation commonly associated with diesel engines is generally harmless. However, in gasoline engines (and sometimes in diesels) detonation explosions can actually dig out pieces of the piston, usually near the edge.

Still another form of heat damage is sometimes called piston seizure. If the engine becomes too hot because of extreme operating conditions, a coolant failure, or uneven combustion, the pistons can expand enough to press against the cylinder wall. Seizure occurs when the piston actually sticks (a rare occurrence in passenger car engines). However, even if the pistons do not literally seize, considerable scraping and scarring can result from direct, unlubricated contact between the piston and cylinder wall.

Stress/Mechanical Damage

Several kinds of mechanical damage are possible. If older, high mileage engines are operated at high speeds, it is possible to: (1) crack the pistons, (2) break or bend a rod, or (3) work loose a piston lock ring. If an engine is old enough, this kind of damage can occur even when the engine is operated at normal speeds. Similar damage can also take place in a rebuilt engine if there is misalignment of some other vital component, such as a piston rod.

PISTON ASSEMBLY/SERVICE

Normal piston wear is considered to be:

1. A slightly loose top ring due to ring groove wear.
2. Horizontal scoring (parallel to the piston head) on the side of the skirt.

Pistons exhibiting these signs of wear can be refurbished by persons skilled in the craft. Pistons with cracks, holes, or other obvious signs of damage or wear are thrown away. As a practical matter, even pistons that can be rebuilt are often traded in on new or previously rebuilt pistons. The old pistons are sent to a specialty automotive machine shop.

Usually, three types of service are performed on pistons and related components:

1. Piston grooves, if worn, may be machined slightly larger and a spacer ring fitted alongside the new ring.
2. The skirt can be expanded, if it has been compressed to less than standard size. Contrary to what the name says, expansion does not mean placing a tool inside the lower part of the piston and pushing out. Instead, a raised pattern is embossed on the outer surface of the skirt. This pattern, which is applied with a knurling device, effectively increases the diameter of the skirt.
3. The third type of service is not performed on the piston itself, but on associated components. Piston pins and locks can be replaced if the piston is not otherwise damaged. Piston rods can be straightened or replaced. The entire assembly can be balanced.

RELATED SERVICE

Many kinds of major service are related. For instance, repairs to or replacement of piston-related components are usually associated with a major overhaul including new rod and crankshaft main bearings. These topics are discussed in other major services chapters.

SPECIAL TIPS

1. If a visual inspection shows a piston to appear mechanically sound, the next step is to measure it for wear. Figure 26–2 illustrates the procedure recommended by GM. Readings obtained in this manner are compared to the manufacturer's specifications to determine whether to reuse the piston, refinish it, or throw it away.
2. Figure 26–3 shows the order recommended by GM for installing compression and oil rings on a piston.

3. A certain space should exist between the ends of the rings after they are installed in the cylinder. Figure 26–4 illustrates the procedure for checking ring gap.

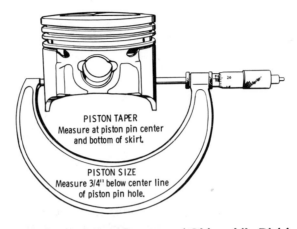

FIGURE 26–2. **Measuring a piston (*Courtesy of Oldsmobile Division, General Motors Corporation*).**

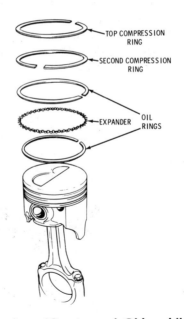

FIGURE 26–3. **Piston rings (*Courtesy of Oldsmobile Division, General Motors Corporation*).**

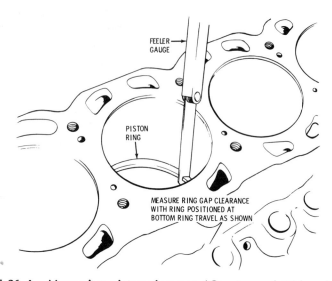

FIGURE 26-4. **Measuring piston ring gap (***Courtesy of Oldsmobile Division, General Motors Corporation***).**

4. Too much side clearance or space in the ring groove can cause piston as well as cylinder damage. If the side clearance is excessive, but still within certain tolerances, it is sometimes possible to reclaim a piston by resizing the groove and installing a spacer below the ring. Figure 26-5 shows how to check the side clearance with a feeler gauge.

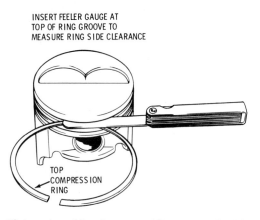

FIGURE 26-5. **Piston ring side clearance (***Courtesy of Oldsmobile Division, General Motors Corporation***).**

5. If a connecting rod is excessively distorted, it must be discarded. Figures 26-6 and 26-7 show some techniques for judging connecting rod bending and twisting.

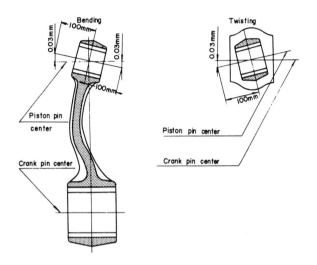

FIGURE 26-6. Measuring bend and twist points (*Courtesy of Nissan Diesel Motors Ltd./Marubeni America Corporation*).

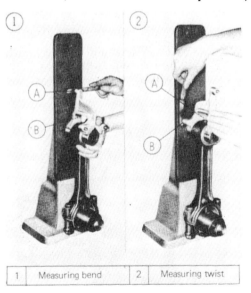

FIGURE 26-7. Tools for measuring bend and twist of connecting rods (*Courtesy of Nissan Diesel Motors Ltd./Marubeni America Corporation*).

6. In addition to checking rod "trueness," it is usually a good idea to find out if the large end of the rod has been worn out-of-round. Figure 26–8 shows how to do this using a dial micrometer.

FIGURE 26–8. Measuring connecting rods at the large end with a dial indicator (*Courtesy of Nissan Diesel Motors Ltd./Marubeni America Corporation*).

7. Particular care must be taken when installing and removing piston rings. Figure 26–9 pictures one type of tool used to minimize ring breakage and piston groove damage. Several varieties of this kind of tool are available.

FIGURE 26–9. Tool for removing and installing rings (*Courtesy of Nissan Diesel Motors Ltd./Marubeni America Corporation*).

8. Figure 26-10 shows a typical ring compressor. It is used to squeeze the rings into the ring grooves so the piston can be more easily inserted into the cylinder bore.

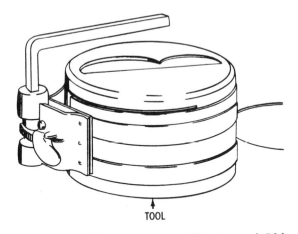

TOOL

FIGURE 26-10.   A piston ring compressor (*Courtesy of Oldsmobile Division, General Motors Corporation*).

9. Figure 26-11 depicts a piston that has just been started into a cylinder. The rings are installed and the compressor is in place. The piston and cylinder have been lightly lubricated. To avoid damage while inserting the piston the rest of the way, a wooden rod is pushed and bumped against the top of the piston.

FIGURE 26-11.   Installing pistons into cylinder block (*Courtesy of Peugeot*).

CHAPTER 27

Lubricating System Service

INTRODUCTION

As noted in Chapter 11, diesel lubricating systems are essentially the same as those used in gasoline powered vehicles. Both types of engines depend on an adequate supply of clean lubricating oils. However, because of extreme pressures and close tolerances, diesels are especially sensitive to any lubrication failure. This is one reason why most diesel manufacturers recommend changing oil and oil filters more frequently than is common in gasoline engines (Figure 27–1).

LUBRICATING SYSTEM PROBLEMS

The keys to lubricating system performance are quantity, cleanliness, and pressure. Each of these factors can be affected by the components in the lubricating system as well as the components being lubricated. For instance, even if the lubricating system is working properly, engine bearing wear will increase the clearance of oil flow passages, thereby reducing oil pressure and quantity. If the wear produces particles and debris, cleanliness is also affected.

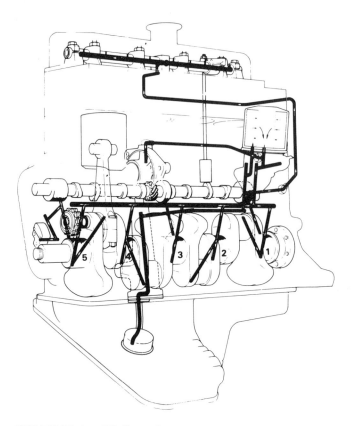

FIGURE 27–1. Oil flow channels (*Courtesy of Peugeot*).

Therefore, if a lubricating problem is suspected, one of the first tasks is to determine which component is the problem. These are the steps in a typical diagnostic procedure.

1. First find out if the oil filter has been changed at the proper interval and if the correct amount of oil is in the reservoir.
2. If the oil pressure light or gauge indicated a problem, check the operation of the unit. This can be done by substituting a gauge of known quality for the suspect unit. Some manufacturers recommend using a special test gauge and comparing the readings to minimum specifications (Figures 27–2 and 27–3).
3. If a reliable gauge shows the oil pressure to be low, drain the oil and remove the oil pan.

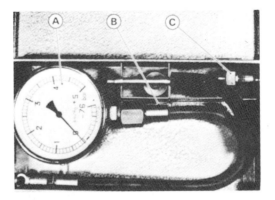

FIGURE 27-2. Oil pressure tester (*Courtesy of Peugeot*).

FIGURE 27-3. Checking the oil pressure (*Courtesy of Volkswagen*).

4. Clean the oil pickup screen (Figure 27-4). A clogged screen will reduce oil flow and pressure.

5. Check the oil pump components for wear or damage. If specifications are not available, make sure: (1) the gear teeth turn freely without excessive backlash, (2) the drive shaft does not have too much end play, and (3) the tips of the teeth are not worn away from the gear housing. If specifications are available, various combinations of these procedures are sometimes followed:

  a. Backlash can be measured by crushing a piece of solder between the gear teeth then measuring the solder with a

FIGURE 27-4. Oil screen and pump (*Courtesy of Nissan Diesel Motors Ltd./Marubeni America Corporation*).

micrometer (Figure 27-5). Backlash is also measured with a feeler gauge placed between the gear teeth (Figure 27-6).

b. Endplay is measured with a straightedge and a feeler gauge (Figure 27-7).

c. Clearance between the tips of the teeth and the gear body is measured with a feeler gauge (Figure 27-8).

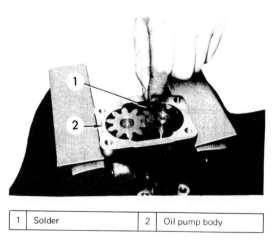

| 1 | Solder | 2 | Oil pump body |

FIGURE 27-5. Measuring backlash with a piece of solder (*Courtesy of Nissan Diesel Motors Ltd./Marubeni America Corporation*).

FIGURE 27-6.   Measuring backlash with a feeler gauge (*Courtesy of Volks-wagen*).

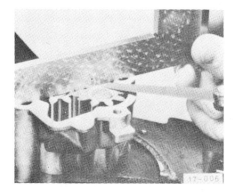

FIGURE 27-7.   Checking end play (*Courtesy of Volkswagen*).

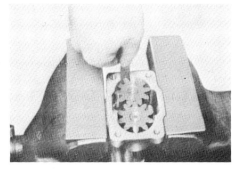

FIGURE 27-8.   Measuring clearance between teeth tips and housing (*Courtesy of Nissan Diesel Motors Ltd./Marubeni America Corporation*).

6. Examine the pressure relief valves.

7. If all these checks do not reveal any problems, the difficulty is likely to be somewhere other than in the lubricating system. At this point, you would probably begin checking the bearing oil clearances.

LUBRICATING SYSTEM SERVICE

Worn or damaged lubricating system parts are usually replaced rather than repaired.

CHAPTER 28

Engine Bearing and Seal Service

INTRODUCTION

Engine seals are relatively inexpensive and are replaced each time an engine is disassembled, even if the seals don't appear visibly damaged. Given the fact that the major expense of tearing the engine down will already have been incurred, it makes sense to do everything necessary before putting the engine back together. New bearings are also routinely installed when the components supported by the bearings are repaired or replaced. This is partly because bearings, like seals, don't cost a great deal of money. It is also because any problems in the components supported by the bearings are likely to be reflected in the bearings themselves. The problems may even appear in the bearings before they do in the supported component. Consequently, whenever a bearing defect is encountered, the underlying cause should be determined to forestall more expensive difficulties somewhere else.

CRANKSHAFT BEARING PROBLEMS

Many of the same problems that cause damage to the crankshaft journals also damage the crankshaft bearings. Some of these conditions include:

1. ***Dirt and Debris.*** There is always a certain amount of loose material floating through the engine in the lubricating oil. Metal dust and occasional small chips and shavings are worn off the oil pump mechanism, the timing gear teeth, and parts of the camshaft and lifters. The smallest particles are normally trapped in the oil filter while the larger, heavier particles usually settle to the bottom of the oil pan. However, if the oil filter becomes too clogged, or if there is any sort of lubrication failure, or if wear between components becomes too severe (which can accelerate as a result of a lubrication problem), particles can get between the bearing and the journal. Some pieces may even become embedded in the soft babbit bearing face. Whenever there is evidence of wear due to dirt or debris, the underlying cause should be determined and corrected before the engine is put back together (Figure 28–1).

Appearance of bearing damaged by *foreign particles in lining.*

Foreign particles in lining create high-spot.

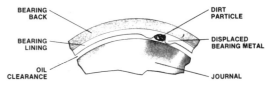

FIGURE 28–1.   Dirt embedded in bearing surface (*Courtesy of Gould, Inc.*).

2. **Lubrication Problems.** A lubrication failure, even if it isn't associated with dirt or debris, can cause damage to the bearings. Excess heat that builds up in the bearing and associated rotating surfaces can weaken the metal of both. Plus, wear will increase at an accelerating pace. Some of the factors that can contribute to a lubrication failure include (Figure 28-2):

 a. A worn oil pump not supplying oil in adequate amounts or at adequate pressures.

 b. A clogged oil filter disrupting the oil flow throughout the engine.

 c. Oil leaks in other parts of the engine, especially around the camshaft bearings. This will disrupt the flow to the engine bearings, which means that the bearings nearest the oil pump may get adequate lubrication while those further away may become starved for oil.

 d. Blocked oil passages are possible for a variety of reasons, but are often due to simple misalignment of the oil supply hole and the oil hole in the bearing.

 e. Reduced oil clearance between the journal and bearing, often the result of an improper replacement bearing selection.

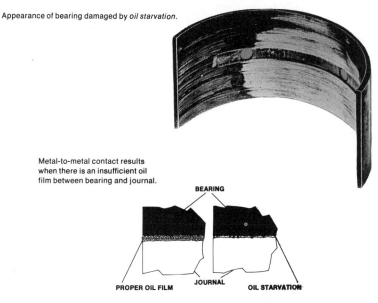

Appearance of bearing damaged by *oil starvation*.

Metal-to-metal contact results when there is an insufficient oil film between bearing and journal.

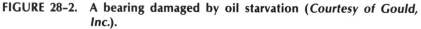

BEARING

PROPER OIL FILM   JOURNAL   OIL STARVATION

FIGURE 28-2. A bearing damaged by oil starvation (*Courtesy of Gould, Inc.*).

3. **Load Related Problems.** Running an engine at excessive speeds or running it so slowly that the engine lugs and bucks strains the bearings and associated rotating surfaces. This strain can have several effects. It can cause the lubricating oil to be squeezed from between the bearings and journals. It can also cause the bearings to flex back and forth. Continued lugging or high speed operation will eventually result not only in bearing failure, but in damage to the components supported by the bearings.

4. **Misaligned Bearings.** If the parts supported by the bearings become bent, warped, or worn in an irregular manner, the bearings will reflect the problem and become misaligned (Figure 28–3).

5. **Corrosion.** As noted in discussing piston problems, low temperature operation allows moisture to condense inside the engine. This moisture combines with combustion by-products to form a corrosive sludge that can damage not only pistons, but bearings and other parts.

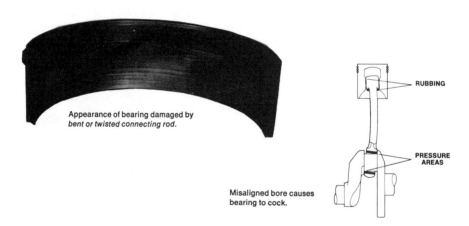

Appearance of bearing damaged by
bent or twisted connecting rod.

RUBBING

PRESSURE
AREAS

Misaligned bore causes
bearing to cock.

**FIGURE 28–3.   A bearing damaged by a bent or twisted connecting rod
(*Courtesy of Gould, Inc.*).**

SEAL PROBLEMS

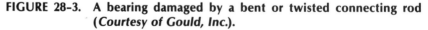

Rubber seals usually leak either because of mechanical stresses or because of a breakdown of the rubber due to age or corrosion. If the problem is due to mechanical stress, the underlying cause should be eliminated before the engine is put back together.

BEARING SERVICE

Similar bearings are used in both gasoline and diesel powered cars and small trucks. However, because of the higher diesel compression ratios, greater stresses are placed on the crankshaft and rod bearings. Consequently, diesel mechanics need to be especially careful in selecting and installing replacement bearings.

Selecting Bearings

The primary objective in selecting replacement bearings is to duplicate, as closely as possible, the original operating condition of the engine. If the crankshaft isn't worn too much, it may be possible to use replacement bearings identical to those installed when the engine was new. If wear exceeds specified limits, special undersize bearings are required. The inside diameter of these bearings is less than the original size.

This procedure can be used to select undersize bearings.

1. Refer to a bearing catalog or a shop manual to find the standard crankshaft diameter of the engine you are working on.

2. Use a micrometer to measure the actual diameter of the worn journals. You will probably find that the journals are worn at least slightly out-of-round. If the wear is within acceptable limits and the journal doesn't need regrinding, use the largest diameter as a basis for determining the correct undersize. If the journal requires grinding, the grinding operation itself will determine the bearing used. Generally, crankshafts are reground to accommodate a specific undersize bearing.

3. If grinding is not required, subtract the actual diameter you measured from the standard diameter listed in the catalog or shop manual. This figure is compared to the standard undersizes available. Normal undersizes are .002", .010", .020", .030", and .060". However, before actually selecting the replacement bearings, it is important to remember that the bearing inside diameter and the journal diameter cannot be exactly the same. A certain amount of clearance is necessary for proper oil flow. Bearing catalogs or shop manuals can be used to find the recommended oil clearance for the engine you are working on (Figure 28-4).

4. In some cases, the exact undersize bearing is not available for a particular journal diameter. In these cases, resizeable bearings can be bored to the diameter required.

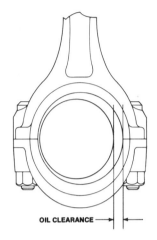

FIGURE 28-4.   Bearing oil clearance (*Courtesy of Gould, Inc.*).

Installing Bearings

In addition to following the factory recommendations when installing bearings, be careful to observe these general guidelines.

    1.   Use a small round wire brush or compressed air to thoroughly clean oil holes and passages in the crankshaft and block (Figures 28-5 and 28-6).

FIGURE 28-5.   Cleaning oil passages in block (*Courtesy of Gould, Inc.*).

FIGURE 28-6.   Cleaning oil passages in crankshaft (*Courtesy of Gould, Inc.*).

2. Wipe off the bearing backs and seats with a clean cloth (Figure 28-7).
3. After the bearings have been snapped in place (Figures 28-8 and 28-9), apply a thin coat of oil to the bearing surface.
4. Make sure the connecting rod offset is correctly positioned (Figure 28-10).

FIGURE 28-7.   Cleaning bearing bores (*Courtesy of Gould, Inc.*).

FIGURE 28-8.   Snapping upper main bearings in place (*Courtesy of Gould, Inc.*).

FIGURE 28-9.   Snapping lower main bearings in bearing caps (*Courtesy of Gould, Inc.*).

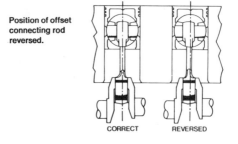

Position of offset connecting rod reversed.

CORRECT   REVERSED

FIGURE 28-10.   Positioning of connecting rod bearing (*Courtesy of Gould, Inc.*).

5. Check for proper shim installation.

6. Make sure the bearing caps are properly positioned.

7. Make sure the locating lugs on the bearings fit into the proper slots.

8. Make sure the oil holes in the bearing and bearing support are properly aligned.

SEAL SERVICE

Seals are normally used to prevent liquids or gases from escaping around the ends of rotating shafts. The air conditioning pump has a seal. So does the water pump. However, the seals normally encountered in major service work are located at both ends of the crankshaft. These seals prevent oil from escaping out of the crankcase. As noted at the first of this chapter, the oil seals are replaced whenever the crankshaft is removed. The exact kind of seal used and the installation procedure vary somewhat from manufacturer to manufacturer, so consult a shop manual before attempting the work on your own (Figures 28–11, 28–12, and 28–13).

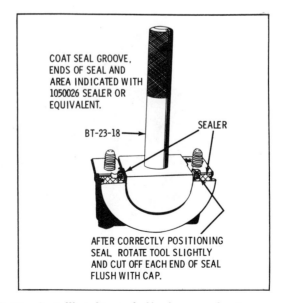

COAT SEAL GROOVE, ENDS OF SEAL AND AREA INDICATED WITH 1050026 SEALER OR EQUIVALENT.

BT-23-18

SEALER

AFTER CORRECTLY POSITIONING SEAL, ROTATE TOOL SLIGHTLY AND CUT OFF EACH END OF SEAL FLUSH WITH CAP.

FIGURE 28–11. Installing lower half of new asbestos rear oil main seal (*Courtesy of General Motors Corporation*).

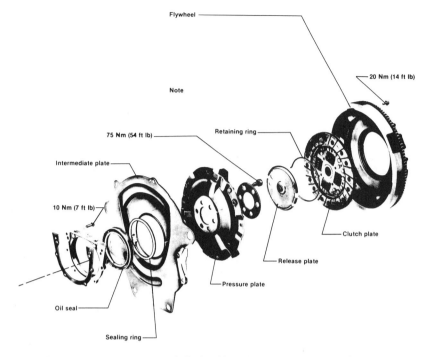

FIGURE 28–12.   Rear crankshaft oil seal (*Courtesy of Volkswagen*).

FIGURE 28–13.   Installing an oil seal (*Courtesy of Gould, Inc.*).

CHAPTER 29

Cylinder Head and Valve Train Service

INTRODUCTION

Valve and valve train work is probably the most commonly performed major service. One reason may be that much of the work can be done without removing the engine from the vehicle; in many cases only the cylinder head needs to be taken off and in some instances not even it needs to be removed. Another more valid reason is the stress these components receive. They are likely to require service earlier than other major components. This is true for diesel as well as gasoline powered engines (Figure 29–1).

VALVE PROBLEMS

Valve problems can be grouped into three major categories: burning, corrosion, and mechanical damage. Often all these problems occur at the same time and are due to related factors (Figure 29–2).

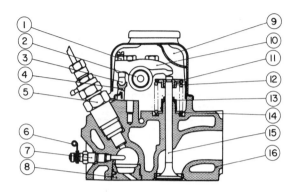

1	Lock nut	9	Rocker cover
2	Rocker shaft	10	Valve rocker
3	Adjusting screw	11	Spring seat
4	Push rod	12	Split collar
5	Nozzle assembly	13	Stem seal
6	Glow plug cable	14	Valve spring
7	Glow plug	15	Valve
8	Combustion chamber	16	Valve seat

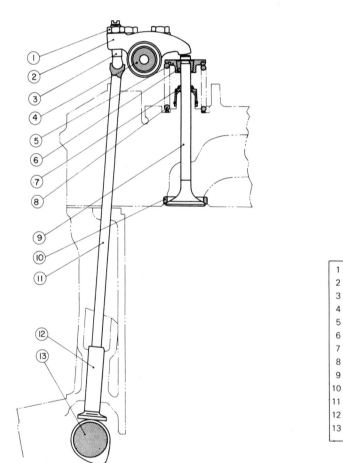

1	Lock nut
2	Valve rocker
3	Adjusting screw
4	Rocker shaft
5	Spring seat
6	Split collar
7	Stem seal
8	Valve spring
9	Valve
10	Valve seat
11	Push rod
12	Valve lifter
13	Camshaft

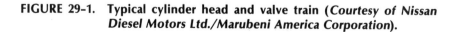

FIGURE 29-1. Typical cylinder head and valve train (*Courtesy of Nissan Diesel Motors Ltd./Marubeni America Corporation*).

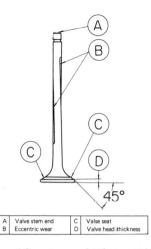

| A | Valve stem end | C | Valse seat |
| B | Eccentric wear | D | Valve head thickness |

FIGURE 29-2.   Valve wear (*Courtesy of Nissan Diesel Motors Ltd./Maru-beni America Corporation*).

Burning

Burned valves (or guttered valves as they are sometimes called) usually result from improper valve seating. If the valve face doesn't fit properly in the valve seat, hot gases are likely to slip by. These gases are so hot that the valve can be literally burned away, starting at the edge where the valve is thinnest and working toward the center. The problem is made worse by the fact that valves are designed to dissipate some of their heat through the valve seat into the cylinder head. If the seating isn't proper, heat transfer cannot take place and the valve becomes that much hotter.

Any number of factors can cause improper valve seating. Carbon deposits that work loose from the combustion chamber can become lodged between the valve face and the valve seat, preventing complete closure. If the valve guide becomes clogged with burned oil deposits, the valve may stick and fail to seat properly. The valve seat can, due to wear or damage, become misaligned with respect to the valve face. If the cylinder head warps, the valve may stick in the valve guide or it may seat at an improper angle.

Corrosion

Corrosion is similar to burning in that the valve gradually deteriorates rather than abruptly breaks (although breakage is likely to accompany extreme burning or corrosion). Corrosion occurs when combustion gases become so hot that chemical action literally eats away the valve material. Such corrosion is temperature-dependent: the hotter the engine, the faster the corrosion proceeds. Chemical corrosion is also likely to be accompanied by physical erosion. Gases tend to move so fast through the ports that the valve material is worn away.

Mechanical Failure

Most of the problems that cause burning and corrosion will eventually wear the valves so much that breakage occurs. The valve might crack; small pieces could break loose; or the entire valve head could fall off into the combustion chamber and pierce the top of the cylinder head.

Breakage can also result from high speed operation (particularly if the valve seating becomes imprecise and uncontrolled). Cracks can occur if the valve temperature changes so rapidly that shock sets in (similar to cracking a cold glass by placing it in hot water).

VALVE TRAIN PROBLEMS

Problems with the camshaft, lifters, push rods, and rocker arm assembly are usually mechanical in nature. These parts will wear due to loss of lubrication or simply as a result of high mileage. Cam lobes, as noted in a previous chapter, can become pitted and chipped. After the hard outer surface is damaged, the wear accelerates, eventually resulting in a loss of cam profile. Damage to the cam lobe will also cause corresponding wear on the lifters, e.g., "cupping" the lifter contact surface. Other types of wear can include bent push rods and a loss of tension in valve springs.

If the wear continues past a certain point, the parts may break. Breakage can also occur as a result of extreme operating conditions (overheating, overrevving, etc.).

CYLINDER HEAD PROBLEMS

The cylinder head contains the top part of the combustion chamber, the intake and exhaust ports, and coolant flow passages. The head also contains the valves. Consequently, the condition of the head can have a considerable affect on engine performance.

Fortunately, the cylinder head, like the block, hardly ever fails so completely that replacement is necessary. Infrequently the head may become warped due to excessive heat or to improperly torqued head bolts. Occasionally the coolant passages may become clogged (Figure 29–3).

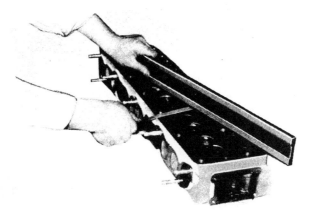

FIGURE 29-3. Checking head warp (*Courtesy of Nissan Diesel Motors Ltd./Marubeni America Corporation*).

The most common problems are wear or damage to the valve guides or valve seats. In some engines these parts are not made from the same material as the head but are fabricated from another metal and press fitted into the head. Not only does this allow more suitable materials to be used, but it also aids repairs or replacement.

Valve guide wear is usually characterized by an increase in the internal bore diameter at both ends of the guide. This lets the valve stem flop around as it travels up and down in the guide. That action, in turn, results in improper seating.

The valve seat simply wears away after repeated sharp contact with the valve face. This prevents the valve face and seat from fitting properly.

VALVE SERVICE

Unless a valve is bent, burned, corroded, or otherwise visibly damaged, it can usually be reconditioned for additional use. A common technique is to regrind the valve face so it makes a proper fit with the valve seat. Following are some tips related to valve grinding (Figure 29-4).

1. Excessive grinding can weaken a valve. As a general rule, the valve margin (area between valve face and head) should never be less than half the margin of a new valve.
2. The valve face in some applications is not designed to fit flush with the valve seat. Instead, the seat and face are ground so that a specified angle (called the interference angle) exists

between the two. The interference angle provides a single line rather than a zone of contact. As a result, valve closing is more immediate and positive.

3.   After a reconditioned valve is inserted into a new or reconditioned valve guide, the valve length is checked. If the valve is too long (and therefore incapable of seating properly), a certain limited amount can be ground off the tip of the stem. If the valve is too short, the valve seat can be ground down (again, within limits) to lower the contact ring.

FIGURE 29-4. Grinding a valve (*Courtesy of Nissan Diesel Motors Ltd./Marubeni America Corporation*).

VALVE SPRING SERVICE

Valve springs can lose tension after a period of use. Both the length and tension should be checked and compared to factory specifications. Use new springs if the old springs fail either test (Figures 29-5 and 29-6).

VALVE GUIDE SERVICE

The internal diameter of an excessively worn valve guide can be reduced by one of two techniques.

1.   The inside can be knurled by cutting a fine spiral groove with a special knife-like tool. After the raised edge of the groove

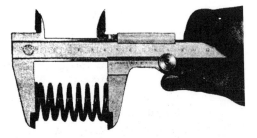

FIGURE 29-5. Checking valve spring length (*Courtesy of Nissan Diesel Motors Ltd./Marubeni America Corporation*).

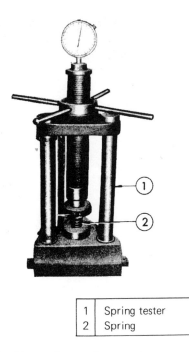

| 1 | Spring tester |
| 2 | Spring |

FIGURE 29-6. Checking valve spring tension (*Courtesy of Nissan Diesel Motors Ltd./Marubeni America Corporation*).

has reduced the internal diameter less than the diameter of the valve stem, the bore is reamed back out to the proper size.

2. A coarse-cut thread can also be tapped into the guide. Then, a brass, spring-like insert is screwed into the thread. After the insert is firmly in place, it is reamed out to the correct diameter (Figures 29–7 and 29–8).

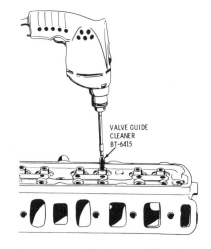

FIGURE 29–7. Cleaning valve guide bores (*Courtesy of General Motors Corporation*).

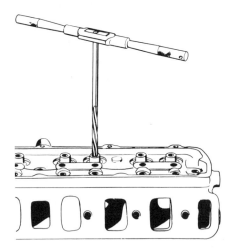

FIGURE 29–8. Reaming valve guide bores (*Courtesy of General Motors Corporation*).

VALVE SEAT SERVICE

Valve seats that are slightly damaged or worn can be reground. The object is to make sure the valve seat is centered (not skewed) with respect to the valve guide as well as to make sure the interference angle corresponds to the factory specifications. The grinding tool is supported in a fixed position so that the cutting stone is centered over the valve guide (Figure 29-9).

In both valve face and valve seat grinding operations, the amount of metal removed is kept to a minimum. If the diameter of the valve is reduced too much and/or the diameter of the valve seat increased excessively, proper valve seating cannot take place (Figure 29-10).

FIGURE 29-9. Grinding a valve seat (*Courtesy of Nissan Diesel Motors Ltd./Marubeni America Corporation*).

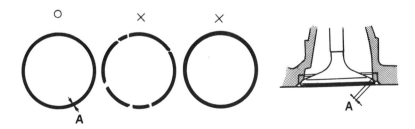

FIGURE 29-10. Checking aspects of valve seating (*Courtesy of Nissan Diesel Motors Ltd./Marubeni America Corporation*).

One way to maintain correct seating is to grind multiple concentric surfaces in the valve seat. Using a combination of top angle and throat angle bevel cuts, the valve seat (which occurs between these cuts) can be effectively raised or lowered. It is something like countersinking a screw, with the valve representing the screw and the seat representing the countersunk portion of the hole.

CYLINDER HEAD SERVICE

In addition to service performed on the valve seat and valve guide, the cylinder head face can be milled level if it is warped. Minor irregularities can be removed by careful application of a fine-cut flat file.

VALVE COMPONENT SERVICE

These components are hardly ever repaired. If inspection proves a part to be defective, it is replaced.

VALVE LASH ADJUSTMENTS

The hydraulic valve systems employed in some cam-in-block engines do not provide mechanisms for adjusting the clearance between the valve stem and rocker arm. Cam-in-head and mechanical cam-in-block systems, however, usually do require periodic valve adjustments. The procedures vary from engine to engine, so consult a factory shop manual before attempting this kind of service (Figure 29-11).

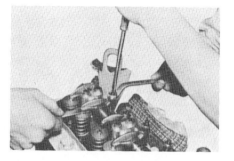

FIGURE 29-11. Adjusting valve clearance (*Courtesy of Nissan Diesel Motors Ltd./Marubeni America Corporation*).

CHAPTER 30

Engine Gasket Service

INTRODUCTION

The primary function of a gasket is to seal off certain areas inside the engine; in other words, to prevent the fluids and gases contained in the engine from escaping to the outside. Gaskets are usually constructed from materials that yield under pressure. That way they can conform to the slight irregularities between mating surfaces and provide a more effective seal. The particular kind of gasket material used is determined by the nature of the application. Some common automotive gasket materials include: rubber, cork, asbestos, paper, and soft metals like copper. Some common applications are the cylinder head gasket, the intake and exhaust manifold gaskets, the valve cover gasket, the water pump gasket, and so on (Figure 30-1).

GASKET PROBLEMS

After continued exposure to water, oil, or fuel, a gasket may deteriorate and lose its ability to provide a proper seal. Gaskets can also fail if the mating surfaces on either side of the gasket are improperly tightened or

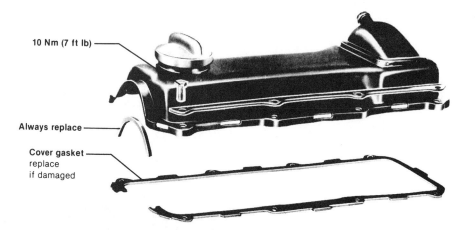

10 Nm (7 ft lb)

Always replace

Cover gasket
replace
if damaged

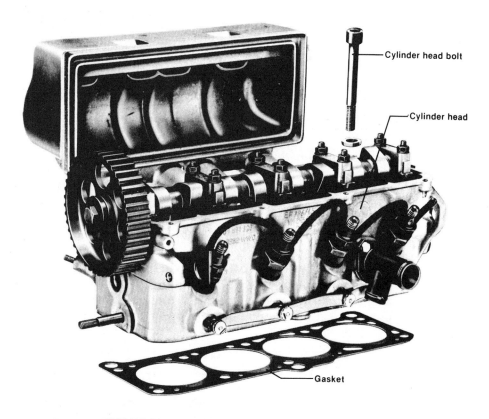

Cylinder head bolt

Cylinder head

Gasket

FIGURE 30-1.   Engine gaskets (*Courtesy of Volkswagen*).

if they become warped. A gasket can likewise become damaged if the mating surfaces separate. For instance, if a leak develops between the exhaust manifold and the head, the hot gases escaping from the leak can burn the exhaust gasket.

GASKET SERVICE

Whenever two mating surfaces are separated, the gasket is often damaged by the separation process. So, most gaskets are replaced as a matter of course whenever they are encountered during a major service project.

COMMON TYPES OF GASKETS

The following paragraphs briefly review some common (but not all of the) gaskets used in both gasoline and diesel powered engines.

Cylinder Head Gaskets

Cylinder head gaskets seal off the high pressure, high temperature gases contained in the combustion chamber. These gaskets are often made from thin pieces of soft metal. Sometimes, tough temperature-resistant asbestos cylinder head gaskets are used. Asbestos gaskets are frequently provided with metal edges, since that is where the pressures are most destructive (Figure 30-2).

Follow these general guidelines when installing cylinder head gaskets.

1. Make sure the head and block mating surfaces are smooth and clean.
2. If the manufacturer recommends using a gasket sealer, apply a thin, even coat on both sides of the gasket.
3. Make sure that the gasket fits the block and that it is installed with the correct side up.
4. Carefully lower the head into position. The gasket is easily damaged if the head is dropped edge first onto the block.
5. Tighten the head bolts in the proper sequence at the correct torque levels. Improper bolt tightening can cause leaks and early gasket failure and can possibly warp the block and/or head.

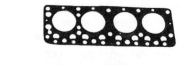

FIGURE 30-2. Four- and six-cylinder head gaskets (*Courtesy of Nissan Diesel Motors Ltd./Marubeni America Corporation*).

Exhaust Gaskets

Exhaust gaskets do not have to withstand high pressures but must resist extreme temperatures. Most exhaust gaskets are made from an asbestos and metallic compound. Installation requirements include clean, smooth mating surfaces and properly torqued attaching screws.

Oil Pan Gaskets and Seals

Oil pan gaskets are usually made from a dense paper compound. Installation requirements again include clean, smooth mating surfaces and evenly torqued screws. Special care should be exercised to avoid distorting the oil pan mating surfaces. The seals at the ends of the oil pan (when used) are usually made from rubberized material (Figure 30-3).

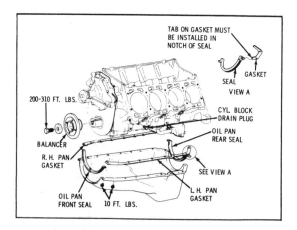

FIGURE 30-3. Oil pan assembly and gaskets (*Courtesy of General Motors Corporation*).

Valve Cover Gaskets

Valve cover gaskets, like oil pan gaskets, do not have to withstand high pressures. However, they can deteriorate from long exposure to the relatively high temperatures encountered under the hood. Valve cover gaskets are often made from thick rubberized materials. Avoid over-tightening the small cap screws that hold the valve cover in place since this can damage the gasket. (Figure 30–4).

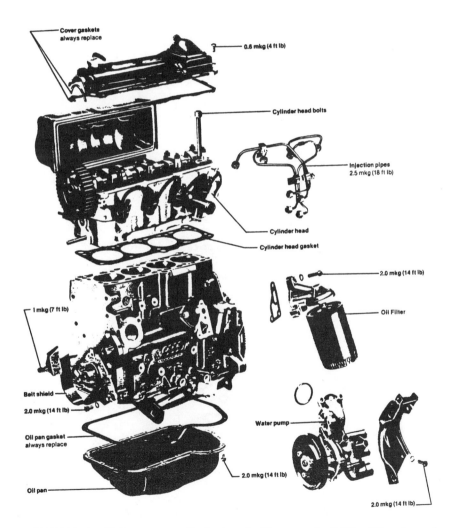

FIGURE 30–4. Valve cover, cylinder head, oil pan, and oil filter (*Courtesy of Volkswagen*).

Front Cover Gasket

Front cover gaskets are usually made from the same paper compounds used in oil pan gaskets. The installation considerations are also similar (Figures 30–5, 30–6, and 30–7).

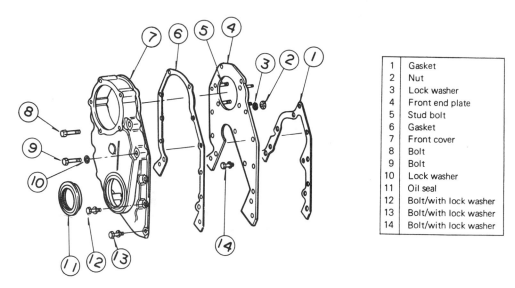

1	Gasket
2	Nut
3	Lock washer
4	Front end plate
5	Stud bolt
6	Gasket
7	Front cover
8	Bolt
9	Bolt
10	Lock washer
11	Oil seal
12	Bolt/with lock washer
13	Bolt/with lock washer
14	Bolt/with lock washer

FIGURE 30–5.   Timing gear case with front cover gaskets and seals (*Courtesy of Nissan Diesel Motors Ltd./Marubeni America Corporation*).

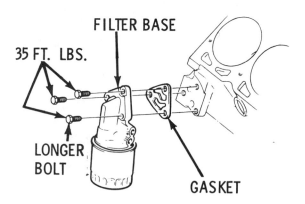

FIGURE 30–6.   Oil filter base and gasket (*Courtesy of General Motors Corporation*).

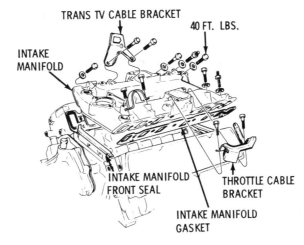

FIGURE 30–7. **Intake manifold and gasket (***Courtesy of General Motors Corporation***).**

CHAPTER 31

Hand Tools

INTRODUCTION

Hand tools, as the name implies, are devices that use human muscle power for operation. Special tools can provide gripping, turning, or cutting power not available to the unassisted body. This chapter is about the common hand tools used for both gasoline engine and diesel work.

SOME REMINDERS

Before reviewing the common hand tools and their functions, it will be helpful to remember some basic facts about tool use.

1. Although everybody repeats this statement, it is still true: Whenever possible, use the correct tool for the job. The wrong tool slows you down and in some cases can cause you to botch a job. For certain tasks, having the right tool is absolutely necessary. For instance, you shouldn't even attempt to

remove a cylinder head unless you have a torque wrench to put it back on.

2. Never use any more force than is necessary when operating a tool. Anyone who has ever twisted a bolt head off doesn't need to be reminded of this rule.

3. Identify your own tools and be careful about either borrowing or lending tools. Since professional mechanics are expected to supply their own hand tools, they are apt to be sensitive about the loss of or damage to expensive equipment.

4. Although some mechanics seem to perform very well with a sloppy tool chest full of dirty tools, it is probably better, at least starting out, to strive for neatness.

5. Although the most expensive, chrome-plated tools may not be worth the money, neither are the cheapest tools. If you are a professional, buy equipment that will last a long time. If you are an amateur, you still ought to get the best tools you can afford. They will help make up for your shortcomings in other areas.

A TYPICAL TOOL BOX

The following paragraphs list the items commonly found in a well-equipped tool box.

Screwdrivers

The two most common types of screwdrivers are the Phillips and slotted-blade kind. They come in all types and sizes; make sure to use big screwdrivers for big screws and little ones for little screws. This may seem obvious, but from an examination of the random methods many people seem to employ when selecting a screwdriver, it really isn't (Figures 31–1 and 31–2).

FIGURE 31–1.   Phillips screwdriver with plastic handle (*Courtesy of New Britain Hand Tools*).

FIGURE 31-2.   Slotted blade or common bit screwdriver (*Courtesy of New Britain Hand Tools*).

Nutdrivers

A nut driver looks like a screwdriver with a socket wrench fitting at the end. Various (small) size sockets can be connected to quickly attach or remove small nuts (Figure 31–3).

FIGURE 31-3.   Nut drivers (*Courtesy of New Britain Hand Tools*).

Pliers

Various kinds of pliers are available: standard gripping pliers, needle-nose pliers, diagonal cutting pliers, bolt cutting pliers, channel lock pliers, etc. If the proper type of pliers is going to be used for the proper jobs, it is not a bad idea to have one each of the major types (Figure 31–4).

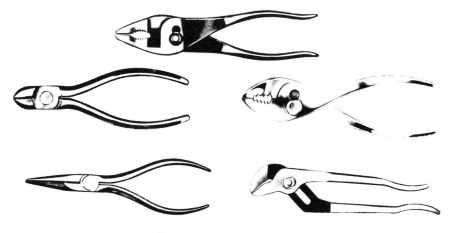

FIGURE 31-4.   Pliers (*Courtesy of New Britain Hand Tools*).

Hammers

Two types of hammers are commonly found in a mechanic's tool box: a steel, ballpeen hammer, and a soft face mallet of some kind. The ball peen should only be used as a last resort. Its indiscriminate use causes more problems than cures (Figure 31–5).

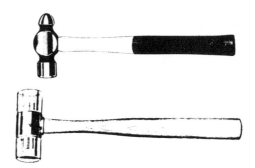

FIGURE 31–5.   Ball peen hammers and soft face hammers (*Courtesy of New Britain Hand Tools*).

Wrenches

Three types of wrenches are commonly used: (1) open end, (2) box end, and (3) combination open/box end wrenches. Each performs the same function; however, the open end is easier to get on and off the nut whereas the box end can sometimes get into a tighter space, plus it has more gripping power. A well-equipped tool box should have a complete range of each kind of wrenches. Given the number of foreign cars on American roads, it is also a good idea to have enough metric size wrenches to fill in the gaps where standard wrenches won't work (Figures 31–6, 31–7, and 31–8).

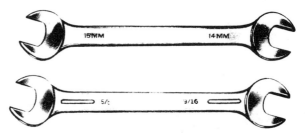

FIGURE 31–6.   Standard open end wrenches and metric open end wrenches (*Courtesy of New Britain Hand Tools*).

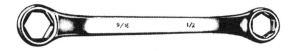

FIGURE 31-7. 6-point box-end wrench (*Courtesy of New Britain Hand Tools*).

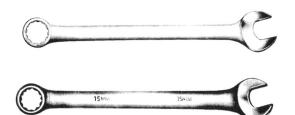

FIGURE 31-8. Combination wrenches (*Courtesy of New Britain Hand Tools*).

Socket Wrenches

Any serious mechanic, whether professional or amateur, should have a socket wrench set with a complete set of sockets and attachments. Work simply goes too slowly without this equipment (Figure 31-9).

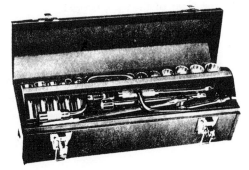

FIGURE 31-9. Socket wrench set (*Courtesy of New Britain Hand Tools*).

Adjustable Wrenches

Adjustable wrenches are handy but don't provide the positive grip necessary for most work. Improperly used adjustable wrenches can quickly round off a nut or bolt head (Figure 31-10).

FIGURE 31–10.   Adjustable wrench (*Courtesy of New Britain Hand Tools*).

Impact Screwdrivers

Impact screwdrivers are used to loosen stubborn fasteners. The tool is held in one hand while the end is struck with a hammer wielded by the other hand. The force of the hammer impact is translated by a cam action into turning power. Different bits (standard screwdrivers, Phillips, etc.) are available. There are occasions when an impact screwdriver is the only tool that can do the job.

Hand-Held Cutting Tools

Mechanics often need to cut or shape metal in various ways. A complete tool box should include:

1. A hacksaw (Figure 31–11).
2. A set of chisels and punches (Figure 31–12).
3. A set of files (Figure 31–13).
4. A tap and die set for cutting threads (Figure 31–14).

FIGURE 31–11.   Hacksaw (*Courtesy of New Britain Hand Tools*).

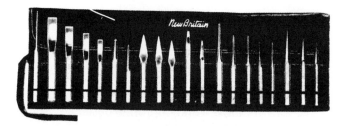
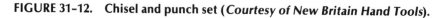

FIGURE 31–12.   Chisel and punch set (*Courtesy of New Britain Hand Tools*).

FIGURE 31-13.   Mill bastard files and round bastard files (*Courtesy of New Britain Hand Tools*).

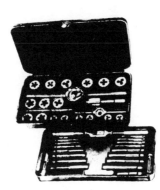

FIGURE 31-14.   Mechanic's tap and die set (*Courtesy of New Britain Hand Tools*).

SPECIAL TOOLS

There are many special tools that make a mechanic's job easier. Some are special tools produced by manufacturers for specific kinds of engines. These tools aren't usually bought by the individual mechanic unless he is working in a very small shop. Many mechanics custom-tailor existing tools to suit their own preferences.

CHAPTER 32

Measuring Tools

INTRODUCTION

Besides needing tools to increase the gripping, turning, or cutting power of the human hand, tools are also required to enhance the judgment of the eye and brain. We must be able to make accurate comparisons between different sizes of things and spaces. We also must be able to say accurately just how large an object or space is or how much something weighs. This is particularly important in engine work because a few thousandths of an inch or a few grams difference between two measurements can have a serious effect on engine performance.

The following paragraphs list some common measuring tools that might be used by mechanics who work on diesel as well as gasoline powered engines.

COMMON TAPE MEASURES AND STRAIGHT RULERS

Most tool boxes contain one or both of these measuring devices. They are adequate for making straight measurements of longer distances. However, most engine related measurements are less than six inches, often less than an inch. Also many engine measurements are not made on flat surfaces along straight lines. The distance along curved surfaces is often measured. Plus, it is often necessary to measure the void or gap between two surfaces. Consequently, more specialized measuring tools are needed for engine work.

FEELER GAUGES

Feeler gauges are used to measure gaps. Three types of feeler gauges are commonly used.

1. **Blade Type** consist of a set of precisely made flat blades, each of which is a specified thickness. Different size blades are inserted into the gap being measured until one fits. The size of the blade that gives the best fit corresponds to the size of the gap (Figure 32–1).

FIGURE 32-1.   All-purpose feeler gauge (*Courtesy of New Britain Hand Tools*).

2. **Wire Type** are made along the same lines as a blade type feeler gauge. However, instead of blades, calibrated wires are used.

3. **Go, No–Go Blade Type** consist of blades thicker at one end than another. This type of feeler gauge is often used where the specification for a gap is given as a high-low range instead of a single figure. The gap is checked by using a feeler gauge blade with the same high-low range. If the smaller end of the blade fits inside the gap, but the larger end is too big, the gap is OK (Figure 32–2).

FIGURE 32-2. Wire and go-no-go feeler gauge set (*Courtesy of New Britain Hand Tools*).

Feeler gauge blades (or wires) are calibrated in inches or metric units. Sometimes both types of units are marked; sometimes reference numbers are used instead of the actual thickness. For instance, a series of blades ranging in thickness from .001″ to .009″ might simply be marked 1 through 9 (Figures 32-3 and 32-4).

FIGURE 32-3. Combination overhead valve feeler gauge set (*Courtesy of New Britain Hand Tools*).

FIGURE 32-4. Piston feeler gauge set (*Courtesy of New Britain Hand Tools*).

CALIPERS

Two types of calipers are occasionally used by engine mechanics.

1. **Transfer Type** calipers resemble well-made tongs with a thumbscrew for adjusting the position of the legs. These calipers have no graduation and are not measuring devices as such. They are used to transfer an otherwise inaccessible dimension to a ruler for measurement. Although not generally believed to be accurate enough for most engine work, transfer calipers are handy for quickly comparing the dimensions of cylindrical objects (like pistons and cylinders) (Figure 32–5)

2. **Vernier Calipers** are precision instruments capable of measuring cylinder and shaft dimensions within a thousandth of an inch. The average automotive technician does not use a vernier caliper as often as his counterpart in a machine shop (Figure 32–6).

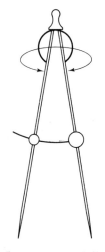

FIGURE 32–5.   Caliper

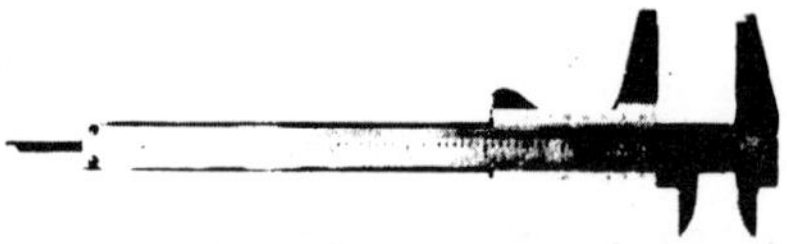

FIGURE 32–6.   Vernier caliper (*Courtesy of New Britain Hand Tools*).

MICROMETERS

Micrometers are more accurate than calipers, but they also cost a great deal more money. Consequently, they are not found in every mechanic's tool box. However, most shops provide micrometers since no quality measurements can be made without them (Figure 32-7).

Micrometers contain two sets of graduations. Coarse readings appear on the sleeve body and fine readings on the thimble. After the micrometer has been adjusted so that the measuring portion fits whatever is being measured, the sleeve and thimble readings are added to get the total readings.

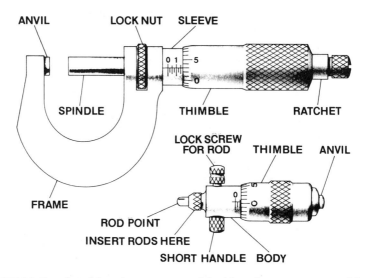

FIGURE 32-7. Outside micrometers and inside micrometers parts identification (*Courtesy of Gould, Inc.*).

Micrometers use a vernier measuring system. Figure 32-8 and the following paragraphs describe how this system works.

1. Turn the thimble until the measuring part of the micrometer fits the span being measured. Never force the thimble, it can damage the micrometer.
2. As the thimble is being screwed in or out, notice how the edge of the thimble moves across the scale on the sleeve body. Every time the thimble makes one revolution, its edge passes over one small division on the sleeve scale.

3. After the thimble adjustments have been completed, determine where the edge of the thimble cuts across the graduations on the sleeve. In the example shown in Figure 32–8, the thimble edge is nearest the 4.00″ mark on the sleeve. Therefore, we can say the coarse reading is 4.00″. The object being measured is actually thicker because there is still some distance between the 4.00″ mark and the edge of the thimble.

4. Observe the graduation marks on the thimble. They are used to subdivide the smallest graduations on the sleeve into even finer readings. In this example, the smallest thimble graduations are at .001″ intervals.

5. Check to see where the horizontal (or revolution) line on the sleeve strikes the scale on the thimble. It is .12″. This figure represents the fine reading; in other words, the distance from the 4.00″ mark to the edge of the thimble. Adding .12″ to 4.00″ gives the total reading of 4.12″, the actual thickness of the object being measured.

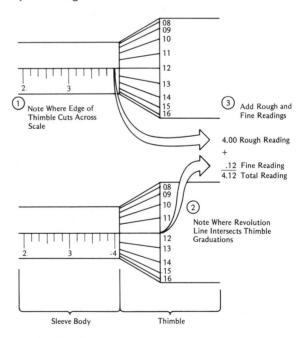

FIGURE 32–8. Reading a micrometer.

Figure 32–9 shows additional tips for reading micrometers. The best way to become proficient is to practice measuring objects of known thickness.

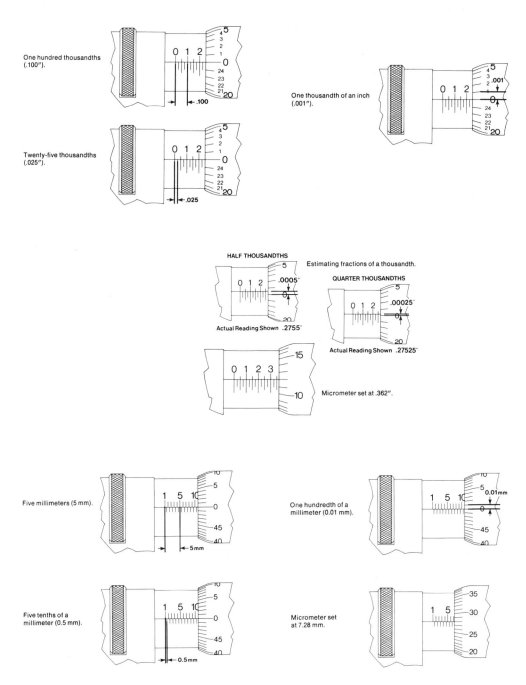

Estimating fractions of a thousandth.

FIGURE 32-9.   Additional micrometer tips (*Courtesy of Gould, Inc.*).

DIFFERENT KINDS OF LENGTH MEASURING SCALES

The three basic kinds of length measuring scales are: fractional, decimal and metric.

Fractional scales divide inches into halves, quarters, sixteenths, thirty-seconds, etc. Most tape measures use fractional scales. So do most common household rulers.

Decimal scales divide inches into tenths, hundredths, and thousandths. Many engine dimensions are given in decimal values, so this scale is one of the most commonly needed. If a decimal ruler is not available, fractions can be converted to decimals by referring to a decimal equivalent table or by dividing the bottom part of the fraction into the upper part. For instance, to convert 1/8th of an inch to a decimal value, 8 is divided into 1. The result is .125 inches.

Metric scales don't use inches at all. Distance is measured in meters, decimeters, centimeters, and millimeters. A meter, which is a little less than a yard, equals ten decimeters. A decimeter equals ten centimeters. A centimeter equals ten millimeters. All measurements in the metric system, including weight and volume measurements, are based on groupings of ten. Since many automobiles, both domestic and foreign made, use metric measurements, a metric scale is often required.

VOLUME MEASUREMENT

Besides needing to measure distances, mechanics often need to measure volume: e.g., the amount of liquid to pour into a container, the volume of a cylinder, etc. Common English or standard units of volume measurement are: quarts, pints, gallons, and cubic inches. Common metric units of volume measurement are liters (approximately equal to quarts), milliliters, and cubic centimeters (cc's). Common measuring devices are graduated cylinders, flasks, and cups.

WEIGHT MEASUREMENT

It is also occasionally necessary to measure weights. Common weight measurement units are pounds, ounces, grams, kilograms, etc. Common weight measurement devices are scales and balances, both the portable and fixed position variety. Many shops have a weight scale of some kind.

TORQUE WRENCHES

Torque wrenches tell a mechanic how much force he is applying to a nut or bolt. The scale on the wrench is calibrated in pound-feet units. Most major service work should not even be attempted without a torque wrench (Figure 32–10).

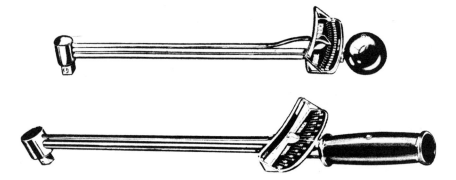

FIGURE 32–10. Torque wrenches (*Courtesy of Gould, Inc.*).

CHAPTER 33

Shop Tools and Major Service Equipment

INTRODUCTION

In addition to the equipment that might be found in an individual mechanic's tool box, a number of other tools are required for professional and efficient engine repair. These large, and/or expensive tools are usually the property of the shop. Most types are suited for both gasoline and diesel engine service.

LIFTING EQUIPMENT

Repairs often proceed more easily if the entire vehicle or a portion of it is lifted off the shop floor. Some of the tools for performing these tasks are pictured in the following illustrations (Figures 33–1, 33–2, 33–3A, 33–3B, 33–4).

FIGURE 33-1.   One-and-three-quarter ton, air-operated end lift (*Courtesy of DeKoven Manufacturing Company*).

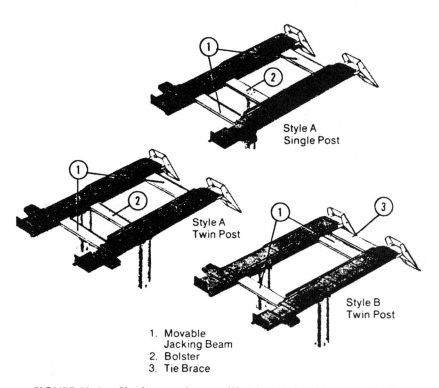

FIGURE 33-2.   Single or twin post lift (*Courtesy of Ammco Tools, Inc.*).

FIGURE 33-3. Portable, lightweight hydraulic garage jack (*Courtesy of DeKoven Manufacturing Company*).

FIGURE 33-4. Jack stands (*Courtesy of DeKoven Manufacturing Company*).

NOTE/CAUTION

Be very, very careful when using any type of lifting equipment. Make sure the lifting point is strong enough to carry the weight. After the vehicle has been raised, make sure it is properly supported. Carelessness or mistakes in this area can lead to fatal surprises.

CLEANING EQUIPMENT

One of the most important aspects of successful engine repair is making sure that the parts going back into the engine are clean. Most shops include one or more of the following types of cleaning devices.

1. ***Immersion Tanks.*** This is one of the most common types of cleaning tools. The simplest version is an old drum cut in half. Other, more expensive variations include agitators and lift-out wire baskets for holding small objects. Whatever the form, the tank is filled with a liquid solvent to loosen or dissolve the dirt, gum, or other deposits that build up on automotive parts. To prevent chemical burns, wear protective clothing, gloves and goggles (Figures 33–5A, 33–5B).

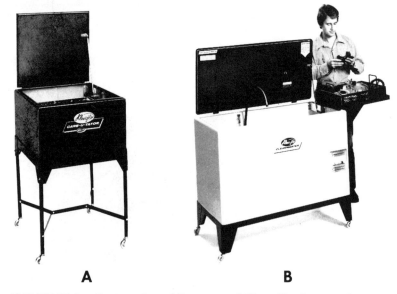

A   B

FIGURE 33-5.   Part washers (*Courtesy of Kleer-Flo Company*).

2. ***Hot Immersion Tanks.*** These tanks are like regular immersion tanks except that the solvent is heated by gas burners or electric coils. Extreme care should be taken when this cleaning method is employed, especially when parts are being lowered into or removed from the tank (Figure 33–6).

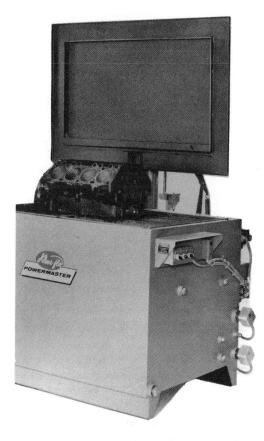

FIGURE 33–6. Gas-operated, heavy duty hot tank cleaner (*Courtesy of Kleer-Flo Company*).

3. ***Steam Cleaners.*** These devices are especially important to used car dealers. A hose hooked up to a steam generating unit can break loose years of grime and dirt from an old engine, making it look virtually like new. Of course, removing dirt makes the mechanic's job easier also (Figure 33–7).

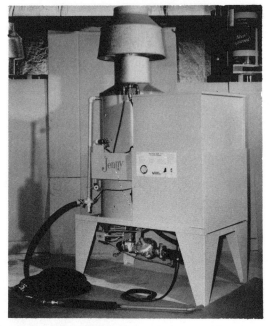

FIGURE 33-7.   Jenny 760 gas-fired steam cleaner (*Courtesy of Jenny Division, Homestead Industries*).

AIR-POWERED HAND TOOLS

Most modern shops have an air compressor. It provides the operating power for a number of tools and devices, including the following.

1. *Impact Wrenches.* These hand-held wrenches use various sizes of sockets for a number of tasks where a great deal of force needs to be applied in a not-too-precise manner: e.g., to loosen nuts and bolts, tighten wheel lugs, etc. However, even where precision isn't required, these tools should be used with care. If not adjusted properly, an air impact wrench can easily twist the head off a nut or bolt (Figures 33-8, 33-9, and 33-10).

2. *Impact Knives/Cutters.* These tools employ a vibrating motion to operate a cutting element. Some uses of air-powered "zip" guns (as they are sometimes called) include shearing off frozen nuts or bolts and cutting sheet metal (Figure 33-11).

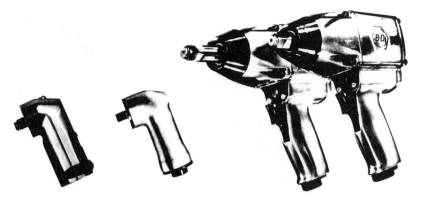

FIGURE 33-8.   Air-operated impact wrenches (*Courtesy of Black & Decker*).

FIGURE 33-9.   An air-operated ⅜" ratchet (*Courtesy of Black Decker*).

FIGURE 33-10.   An electric-operated ¾" impact wrench (*Courtesy of Black & Decker*).

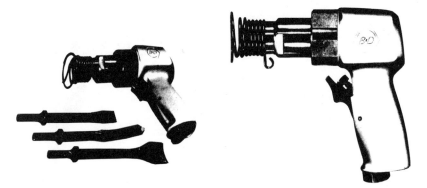

FIGURE 33-11.   Air-operated power chisels (*Courtesy of Black & Decker*).

3. *Air Hoses.* The common air hose and nozzle can be considered an air-powered tool. It is used to clean or dry small parts. The air hose is also the object of horseplay between mechanics in some shops. Even though mildly humorous, games with air hoses should be avoided. Thoughtless "air gooses" have already resulted in too many distended or ruptured lower intestines.

GRINDERS

Most commercial grinders contain two wheels mounted on either end of a common motor shaft. The solid or cutting wheel is used to shape metal parts and custom tailor tools. The wire brush wheel is used to polish metal surfaces. Both wheels, especially the grinding wheel, can produce showers of hot metal filings. Operators must protect themselves by wearing goggles and a shop apron (Figures 33-12 and 33-13).

LUBRICATING TOOLS

Any shop, large or small, will probably have a hand-operated grease gun. Many shops also provide power operated greasing tools (Figure 33-14).

FIGURE 33-12. Deluxe heavy duty valve refacer (*Courtesy of Black & Decker*).

FIGURE 33-13. Bench and hand grinders (*Courtesy of Black & Decker*).

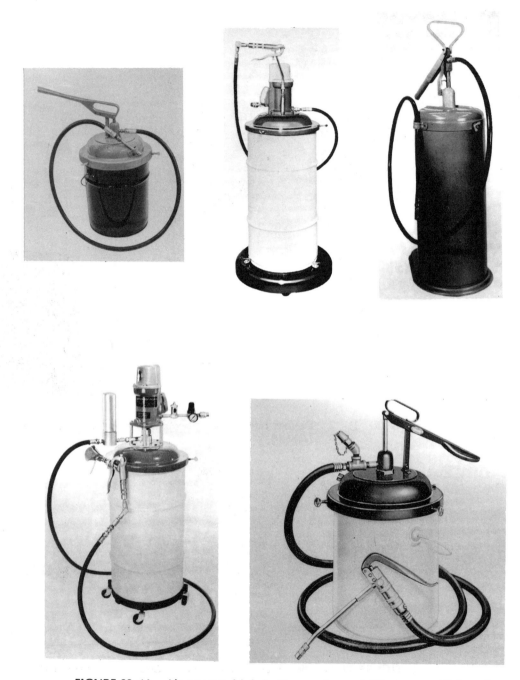

FIGURE 33-14.   Air-operated lubrication equipment (*Courtesy of Balcrank Products Division, Wheelabrator-Frye, Inc.*).

ELECTRIC DRILLS

Most shops also provide electric drills of some kind, either hand-held drills or fixed position drill presses (Figure 33–15).

FIGURE 33–15. **Electric, variable-speed, reversible, ½" drill (*Courtesy of Black & Decker).***

HYDRAULIC PRESSES

A hydraulic press looks something like a drill press. However, instead of a drill bit, the hydraulic press has a solid bar that pushes against an object placed on the press platform. Hydraulic presses are used to straighten bent pieces, remove or install press fit bushings, etc. (Figure 33–16).

TUNE-UP TOOLS

All of the preceding tools can be used equally for work on gasoline or diesel engines. The same is not true for many tune-up tools. Timing lights, oscilloscopes, and tach dwell meters depend on ignition system impulses for the information they provide. Diesel engines do not have an ignition system as such and don't require these tools. Refer to Chapters 19 through 23 for discussion of diesel tune-up procedures (Figure 33–17 and 33–18).

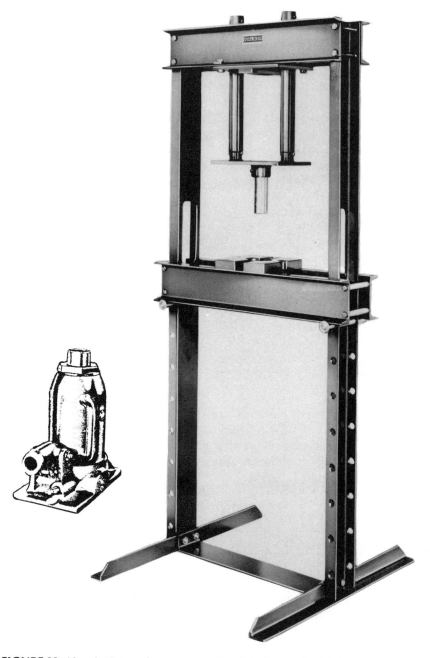

FIGURE 33–16.   A 15-ton shop press and jack to be used with it (*Courtesy of DeKoven Manufacturing Company*).

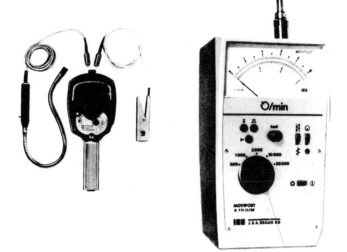

FIGURE 33–17.   Photo-electric cell tachometer (*Courtesy of Peugeot*).

FIGURE 33–18.   Compression gauge (*Courtesy of Peugeot*).

Index